CHANDRA'S COSMOS

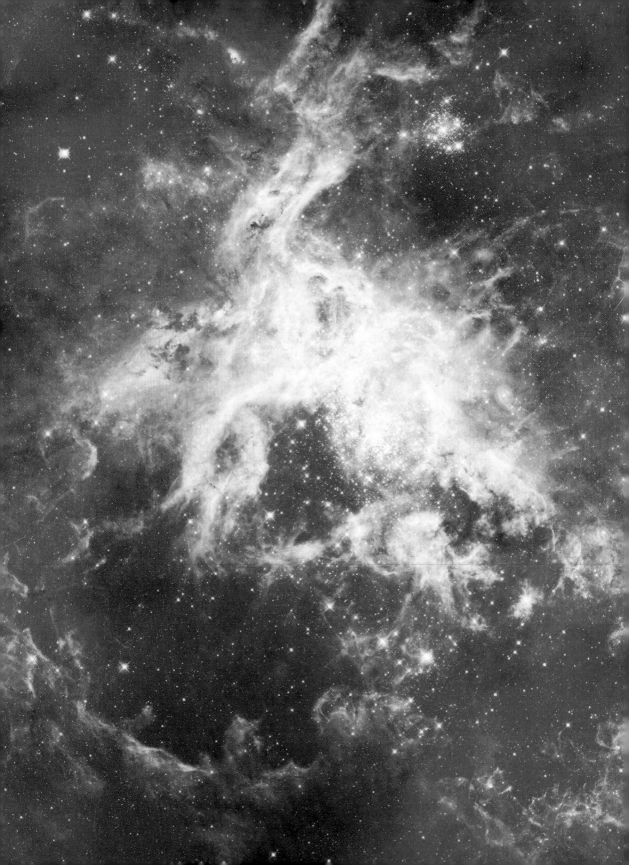

Wallace H. Tucker

CHANDRA'S COSMOS

Dark Matter, Black Holes,

and Other Wonders Revealed by

NASA's Premier X-Ray Observatory

Smithsonian Books

WASHINGTON, DC

This book may be purchased for educational, business, or sales promotional use. For information, please
write: Special Markets Department, Smithsonian Books, P.O. Box 37012, MRC 513, Washington, DC 20013

Published by Smithsonian Books
Director: Carolyn Gleason
Managing Editor: Christina Wiginton
Production Editor: Laura Harger
Edited by Gregory McNamee
Designed and typeset by Jody Billert and Scribe Inc.
Indexed by Clive Pyne

Library of Congress Cataloging-in-Publication Data
Names: Tucker, Wallace H.
Title: Chandra's cosmos : dark matter, black holes, and other wonders revealed
by NASA's premier X-ray observatory / Wallace H. Tucker.
Description: Washington, DC : Smithsonian Books, [2017] | Includes bibliographical references and index.
Identifiers: LCCN 2016016560 | ISBN 9781588345875
Subjects: LCSH: Chandra X-ray Observatory (U.S.) | X-ray astronomy.
Classification: LCC QB472.T8114 2017 | DDC 523.1—dc23
LC record available at https://lccn.loc.gov/2016016560

Manufactured in China through Oceanic Graphic Printing, not at government expense
21 20 19 18 17   5 4 3 2 1

For permission to reproduce illustrations appearing in this book, please correspond directly with
the owners of the works, as seen at the back of the book. Smithsonian Books does not retain
reproduction rights for these images individually or maintain a file of addresses for sources.

Title pages: 30 Doradus, a.k.a. the Tarantula Nebula. See pages 234–35.
Endsheets: Cygnus A (colorized). See page 100.

CONTENTS

1 | IC 443 SUPERNOVA REMNANT

A wide-field view of a supernova remnant (SNR) made with X-ray (blue; from Chandra and ROSAT, the Roentgensatellite), radio (green; from the Jansky Very Large Array), and optical (red; from the Sloan Digitized Sky Survey) observations. Embedded near IC 443's lower edge is neutron star J0617, spewing out a cometlike wake of high-energy particles as it races through the SNR's multimillion-degree Celsius gas.

INTRODUCTION

You will behold through the
telescope a host of other stars,
which escape the unassisted sight,
so numerous as to be almost beyond belief.

Galileo, *The Starry Messenger*, 1610

COOL STORIES FROM
THE HOT UNIVERSE

In Florence, Italy, in the year 1609, the world changed. Using a small telescope, Galileo Galilei proved that Earth is not distinct from the universe but part of it, and that there is much more to the universe than we see with the naked eye. In the twentieth century, astronomers made another revolutionary discovery: Optical telescopes reveal only a portion of the universe. Telescopes sensitive to invisible wavelengths of light have detected microwave radiation from the Big Bang, infrared radiation from protoplanetary disks around stars, and X-rays from matter swirling into black holes.

On July 23, 1999, NASA launched the Chandra X-Ray Observatory aboard the space shuttle *Columbia*. A telescope designed to detect X-ray emissions from extraordinarily hot regions of the universe—exploded stars, galaxy clusters, and matter around black holes—Chandra now orbits above Earth's X-ray-absorbing atmosphere at an altitude of up to 86,500 miles. The Smithsonian's Astrophysical Observatory in Cambridge, Massachusetts, is the site of the Chandra X-Ray

Center, which operates the satellite, processes the data it collects, and distributes it to scientists around the world for analysis.

Since its launch, Chandra has given us a view of the universe that is largely hidden from telescopes sensitive only to visible light. It is a universe of violent and extreme environments, such as intense gravitational and magnetic fields around black holes, supernova shock waves, and titanic collisions between clusters of galaxies.

X-ray astronomy is one of the newest fields of astronomy. Earth's atmosphere absorbs X-rays so strongly that it is impossible to observe them with ground-based telescopes. There is another problem: An X-ray photon has about a thousand times more energy than an optical photon, and because of this high energy, X-rays are difficult to focus onto a detector. X-ray telescopes therefore require extremely smooth mirrors, and Chandra's mirrors are the smoothest ever made for any telescope.

As an X-ray telescope, Chandra collects and images X-rays that it detects from cosmic sources. In essence, Chandra is making the first ever high-resolution X-ray pictures of cosmic X-ray machines, images of a quality equivalent to those of large optical telescopes. This unique capability has transformed twenty-first-century astrophysics.

Chandra is part of a larger effort in which, for the first time, it is possible to make high-resolution images of star clusters, galaxies, and other cosmic phenomena with powerful telescopes tuned to many wavelengths, from radio to X-ray, including the infrared and visible bands in between. Chandra's synergy with the Hubble Space Telescope, which can observe near-infrared, optical, and ultraviolet radiation, and the Spitzer Space Telescope, which detects infrared radiation, along with many other telescopes, is providing a much more complete view of the evolution of stars and galaxies as well as the extreme physics associated with black holes and cosmic explosions.

Since the launch of Chandra, as science spokesman for the Chandra X-Ray Center, I have been privileged to enjoy a front-row seat to the observatory's exploration of the universe. During this time, I have been profoundly impressed by the work of a growing community of ingenious scientists who have used Chandra's unique and critical capabilities in the quest to understand our place in the universe and the fundamental laws that govern our existence.

My role has been primarily to explain or document some of the successes in this quest, and from time to time I have been asked to put together lists of "top ten" Chandra stories. On one of these occasions, a colleague suggested that

I should prepare a list of Chandra's coolest discoveries. Not being an especially cool person—at least I don't think I'm cool, and, as I understand it, if you're cool, you know it—I was a little perplexed. My reaction confirmed one of Malcolm Gladwell's rules of cool: "It can only be observed by those who are themselves cool" (1997).

As I pondered what might constitute cool discoveries, I came across these words of wisdom from best-selling author Stephen King: "The meaning of cool is beyond definition . . . beyond modification. It just is, man" (2007). Not much help there. The words of William Gibson, prophet of the cyberpunk subgenre of noir fiction, seemed more appropriate: "Secrets are the very root of cool" (2007). This fits in nicely with something an unquestionably cool scientist, Albert Einstein, once said: "The most beautiful experience we can have is the mysterious. It is the fundamental emotion that stands at the cradle of true art and true science" (1949).

Secrets and mysteries. I have selected the interconnected stories presented here with these criteria of "cool" in mind. They fall into three broad categories, which we will explore in detail in the pages that follow: the big, the bad, and the beautiful.

The big stories deal with the big picture: What is the universe made of? How did it evolve into its present state? What does the future hold? The past decade has seen revolutionary progress toward finding the answers to these questions. Based upon a broad suite of astronomical observations made with microwave, optical, and X-ray observatories, a consensus has emerged: Only about 5 percent of the matter in the universe is normal matter consisting of protons, neutrons, and electrons; 25 percent is dark matter; and 70 percent is thought to be dark energy, a mysterious force that causes the observed accelerating expansion of the universe. These numbers convey a sobering message: 95 percent of the universe is "dark," a label chosen to express our inability to directly detect or identify it.

We know that dark matter plays a major role in how galaxies form. In the past decade, astrophysicists have found that another type of darkness—black holes—may be critical to the evolution of galaxies. Black holes are the ultimate bad guys of the cosmos. The forces associated with their immense gravitational fields can tear apart anything that ventures too near, be it stars, planets, or atoms. A substantial body of evidence shows that a supermassive black hole with a mass of several million Suns lies at the center of most, if not all, galaxies. Furthermore, the growth of such supermassive black holes appears to be closely connected to the growth of galaxies as a whole.

On a stellar scale, black holes are formed by the collapse of extremely massive stars. This collapse occurs when the nuclear reactions in the center of a star can no longer support the enormous weight bearing down on them from the rest of the star. A star less massive than about eight times the mass of the Sun reaches equilibrium by expelling its outer layers and settling down to become a dense white dwarf star. Stars with larger masses, however, do not leave the stage quietly (figure 1, page vi, and figure 2). Their weight is too much for equilibrium to hold; the central regions collapse to form a neutron star or black hole, and most of the star is blasted away in one of the most violent events in nature, a supernova. If a supernova occurred within a dozen or so light-years of our home planet, it would in all probability extinguish most life on Earth. Whether supernovas have exterminated other civilizations in our galaxy is unknown, but that is certainly a possibility. What *is* known is that supernovas disperse planet-building and life-giving elements manufactured in their interiors—elements such as carbon, nitrogen, oxygen, silicon, calcium, and iron—over regions thousands of light-years in diameter.

Their effects don't end there. If the expanding supernova shock wave slams into a dormant cloud of dust and gas, the impact can trigger the collapse of clumps of gas. A million years later, the sky will blaze with colorful, utterly beautiful lights from a new generation of stars, and possibly new planets and new civilizations. Cool indeed.

2 | CASSIOPEIA A SUPERNOVA REMNANT

A Chandra X-ray image of this remnant, called Cas A for short. Red, green, and blue regions are where low-, medium-, and high-energy X-rays, respectively, are most intense. The red material on the left outer edge is enriched in iron; the bright greenish-white region on the lower left is enriched in silicon and sulfur. In the blue region at right, low- and medium-energy X-rays are filtered out by a cloud of dust and gas. (Note: In this image and many that follow, green colors are not readily apparent. Green, normally used to depict wavelengths between red and blue, does not stand out as much as the latter colors.)

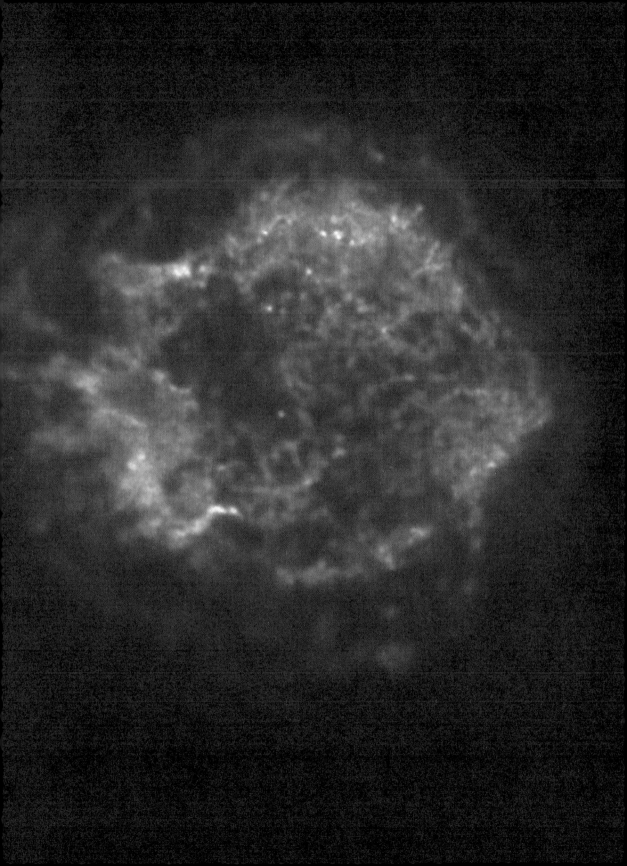

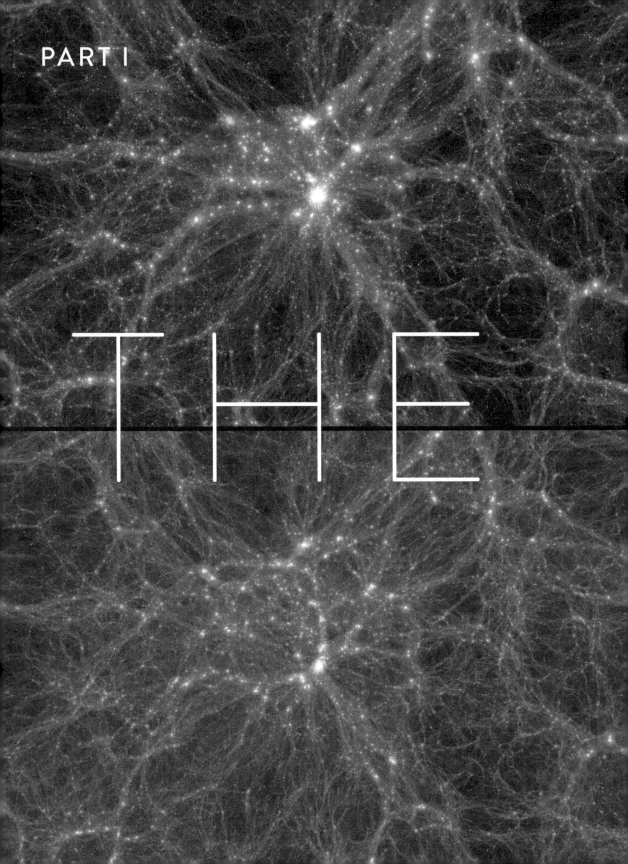

PART I

THE

BIG

3 | COSMIC WEB

Four snapshots from the Illustris project computer simulation, which models the cosmic structures produced by the evolution of dark matter, gas, and galaxies—the framework of the large scale of the universe. These images from the simulation show the distribution of dark matter in the universe as it evolved after the Big Bang: from 1.5 billion years after the Big Bang (top left), 3.3 billion years afterward (lower left), 5.9 billion years afterward (top right), and up to 13.8 billion years afterward, the present time (lower right).

4 | ABELL 1689

A composite image using data from two separate telescopes, showing this enormous cluster of galaxies approximately 2.3 billion light-years from Earth. Purple indicates hundred-million-degree Celsius gas that the Chandra X-Ray Observatory detected in the cluster. Yellow indicates galaxies from optical data gathered by the Hubble Space Telescope.

Clusters of galaxies are far more
numerous than previously thought and . . .
may be the fundamental condensations
of matter in the universe.

George Abell, "The Distribution of Rich Clusters of Galaxies," 1958

GALAXY CLUSTERS

Long before it was understood what spiral and elliptical nebulas were, astronomers—especially Charles Messier (1730–1817), a Frenchman who catalogued them so that they wouldn't confuse his search for comets—noted that they are not distributed uniformly across the sky. Rather, they tend to be concentrated into groups of a few and clusters of hundreds or more, a property noticed and remarked upon by British astronomer Wilhelm Herschel (1738–1822), who presciently noted that the system we inhabit is likely very similar to spiral nebulas. A century and a half later, in the 1920s, studies of groups of clusters of galaxies were revolutionized by Edwin Hubble's demonstration that spiral and elliptical nebulas are galaxies like the Milky Way, located at distances of millions of light-years, which implied that these clusters of galaxies are in fact systems of enormous size.

The tendency of galaxies to cluster together has been confirmed as larger samples of galaxies have been compiled. The first modern catalog of galaxy clusters, published by UCLA astronomer George Abell (1927–85) in 1958, contained 2,712 rich clusters of galaxies found by the National Geographic Society–Palomar Observatory Sky Survey. A *rich cluster* is defined as one containing more than fifty galaxies. New surveys have mapped the distribution galaxies out to distances of several billion light-years. The most comprehensive of these is the Sloan Digital Sky Survey (SDSS). This survey has used the Sloan Foundation's dedicated 2.5-meter optical telescope at Apache Point Observatory in New Mexico to obtain deep, multicolor images covering more than a quarter of the sky and to create three-dimensional maps containing more than 1.2 million galaxies (see, for example, figure 5).

These large-scale maps of the universe, when projected into two dimensions, resemble road maps. Galaxies line up in filaments that crisscross intergalactic space like superhighways. Connecting these are smaller roads, and between these are regions of low population density: the cosmic countryside. At the crossroads, where multiple filaments converge, are clusters of galaxies, the cosmic megacities (figure 4, page 8, and figure 6).

The size of these clusters is daunting. Light, which travels at 186,000 miles per second, takes a little longer than a second to reach Earth from the Moon, and eight minutes to reach Earth from the Sun. Light from the center of our Milky Way galaxy must make a journey of twenty-five thousand years to reach us. Even that is fast compared to the time required for light to cross a galaxy cluster: about 2 million years.

Clusters are the largest gravitationally bound bodies in the universe. The roadlike filaments may be larger in sheer size, but they are not coherent bodies held together by gravity. Like metropolises on Earth, cosmic megacities do not originate fully grown but are assembled from smaller clusters over billions of years. Unlike on Earth, where economic forces strongly influence the formation of cities, gravity is the principal driving force in the creation of galaxy clusters. As astronomers have recently learned, the effects of gravity on the evolution of a cluster can manifest in unexpected ways. The space between the "urban" cores, or galaxies, in a cluster is not empty. It is filled with diffuse gas left over from the formation of galaxies, heated to tens of millions of degrees by the slow gravitational collapse of the cluster. When gravity began to pull galaxies and the clouds in which they are embedded together to form clusters about 11 or 12 billion years ago, the slow collapse heated the intergalactic gas to tens of millions of degrees. This enormous gas cloud contains about five times the mass of all the stars in all the galaxies in the cluster, and it is bound by the gravity of an even more massive

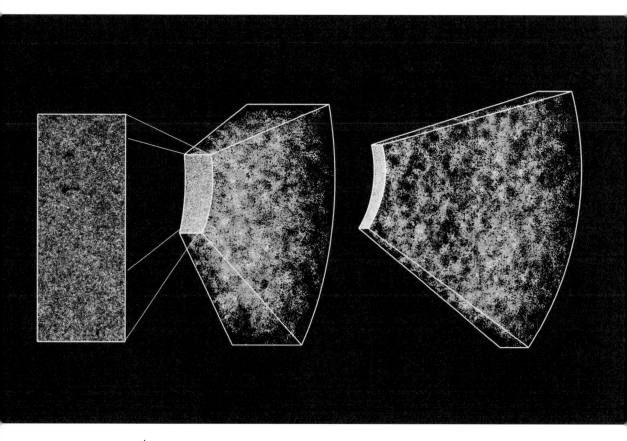

5 | SLOAN DIGITAL SKY SURVEY III MAP

A section of the three-dimensional map constructed by the Sloan Digital Sky Survey III, which used the Sloan Foundation's dedicated 2.5-meter optical telescope at Apache Point Observatory in New Mexico. The rectangle at left is a cutout of 1,000 square degrees in the sky. It holds about 120,000 galaxies (ca. 10 percent of the total survey). Measurements of the optical spectrum of each galaxy—each represented by a dot in this cutout—make it possible to determine the distance to each galaxy and to change the two-dimensional image into a three-dimensional map (at center and right), permitting us to look 7 billion years into the past. Bright regions of this map represent parts of the universe with more galaxies and thus more mass, which exert excess gravitational pull.

cloud of dark matter. Because the hot gas clouds—which optical telescopes cannot see—are confined to the clusters and radiate away their energy very slowly, they can preserve a record of much of the activity in the clusters over the past few billion years. The gas retains the elements injected into it by past supernova explosions in the cluster galaxies, as well as the energy from these and other explosions.

The properties of galaxy clusters place them at another crossroads: the transition between astrophysics and cosmology, the study of the universe at large. The abundance and sizes of clusters bear imprints of the conditions in the primordial background gas from which they emerged. Because of the vast reservoir of gas and the nearly closed-box environment, most of what happens in the cluster stays in the cluster. This makes clusters ideal laboratories in which to study both processes operating during the formation and evolution of galaxies and the supermassive black holes they contain.

In a manner similar to archaeologists, who unearth the past by studying artifacts buried beneath cities and other sites, astronomers are using Chandra and other orbiting X-ray telescopes to "excavate" the rich treasure trove of relics present in galaxy clusters and to piece together their past. Some of this history is relatively recent, and it tells us about the growth of the most extreme objects in the universe: supermassive black holes embedded in giant galaxies in the clusters. Some of this history, however, goes back a very long way, more than 10 billion years, to the beginning of the universe.

6 | CL J1001+0220 GALAXY CLUSTER

A composite of this concentration of galaxies and hot gas 11 billion light-years from Earth. It is the most distant and youngest galaxy cluster detected in X-rays to date. This image shows X-radiation detected by Chandra (purple), infrared emission observed by the European Space Observatory's UltraVISTA survey (red, green, and blue), and radio emission detected by the Atacama Large Millimeter/Submillimeter Array (ALMA; green).

7 | COMA GALAXY CLUSTER

A composite image of the Coma galaxy cluster, about 321 million light-years from Earth. This image was processed to bring out the cluster's large arms of hot gas, which probably formed when the gas was stripped out of smaller subclusters of galaxies as they merged with Coma. Chandra data are in pink, and optical data from the Sloan Digital Sky Survey are in white and blue. The image extends about 2 million light-years on each side.

Eliminate all other factors, and
the one which remains must be the truth.

Arthur Conan Doyle, *The Sign of the Four*, 1890

THE EVIDENCE FOR DARK MATTER

For a long time now, astronomers have used the type of logical reasoning made famous by Arthur Conan Doyle's Sherlock Holmes to deepen our understanding of the universe. Ever since the phenomenal success of Isaac Newton in explaining the motion of the planets with his theory of gravity and laws of motion in 1687, unseen matter has been invoked to explain puzzling observations of cosmic bodies. The anomalous motion of Uranus led astronomers to suggest that an unseen planet existed, and a few years later, in 1846, Neptune was discovered. This procedure is still the primary method used to discover planets orbiting stars. A similar line of reasoning led to the detection in 1862 of a new type of object, a faint white dwarf, which came to be known as Sirius B, in orbit around the bright star Sirius.

Nature does not always reveal her secrets so readily. An attempt to explain anomalies in the motion of Mercury as the effects of a hitherto unknown planet called Vulcan did not succeed. The solution turned out to be Einstein's theory of general relativity, which modified Newton's theory.

Today, astronomers are faced with a similar, though much more severe, problem: dark matter. Simply stated, dark matter is matter that cannot be seen with any type of telescope, but it can be detected through its gravitational effects. These effects are observed as peculiarities in the orbits of stars and clouds of hot gas in galaxies and galaxy clusters. Most astronomers think that all other possibilities have been eliminated and have come to the humbling conclusion that most of the matter in the universe, approximately 80 percent, is dark matter—"humbling" because they do not know what it is. The two known types of dark matter, neutrinos and black holes, are thought to be only a minor portion of the overall dark matter budget.

Unlike the case of Uranus, where the gravity of Neptune adds a fraction of a percent to the gravitational force acting on Uranus, the extra force needed in the cases we will consider here is several hundred percent. It is no exaggeration to say that solving the dark matter problem will require a profound change in our understanding of the universe. It was research on galaxy clusters that initially brought the dark matter problem to the attention of the scientific world—yet this research was largely ignored for almost half a century. It is likely that the study of clusters will be one of the keys to the dark matter solution.

Just a few years after Edwin Hubble and his colleagues showed that spiral and elliptical nebulas are galaxies containing billions and even trillions of stars, he and Milt Humason measured the random motions of galaxies and found that galaxies in clusters had very large random velocities. Fritz Zwicky used this information to show that the total cluster masses for the Coma and Virgo clusters were enormous as well (1933; 1937).

The basic method for determining the mass of an object is to measure the acceleration due to the gravity of matter near the object. For example, if we know the distance of a planet from the Sun and the time that the planet takes to complete one orbit, we can compute its orbital speed and centripetal acceleration. Balancing this data with the gravitational force, which depends on the mass and the radius, gives the mass. This is straightforward enough in our own solar system, where radar ranging can be used to measure distances precisely and the time for an orbit is measured in months for the inner planets and years for the outer ones. For galaxies and clusters of galaxies that are millions of light-years away and make one rotation once every few hundred million years, the process gets a little more complicated, but the basic physics is the same.

Zwicky estimated that ten to a hundred times more matter than can be detected in stars is needed to keep a galaxy cluster from flying apart. He coined the term *dark matter*, suggesting that the unknown matter possibly exists in the form of hot gas undetectable with optical telescopes.

Indeed, in the past three decades, X-ray telescopes have discovered vast clouds of multimillion-degree gas in clusters of galaxies. X-ray observations can measure the mass of the hot gas. They have shown, for example, that the mass of hot gas is about five times that contained in all the trillions of stars in the Coma cluster (figure 7, page 14). That's a lot of matter, but it is not enough. An X-ray telescope can also measure the variation of temperature, density, and pressure in the hot gas from the inner parts to the outside. Pressure falls off with distance in a way similar to the pressure in Earth's atmosphere, which is confined to Earth by its gravitational force. Using the same basic reasoning, the gravitational force needed to confine hot gas to the Coma cluster can be computed. This calculation shows that the mass of the dark matter must be about five to six times greater than the mass of all the stars and gas in the cluster. Adam Mantz of the University of Chicago and his colleagues have confirmed and refined this estimate by using Chandra data from a sample of forty galaxy clusters (Mantz et al. 2014). They found that the mass of dark matter is 5.7 and 7.6 times greater than the amount of baryonic matter in the stars and gas in these clusters. (Physicists call matter composed of protons and neutrons baryonic matter, to distinguish it from electrons and neutrinos, which are called leptonic matter, and from dark matter, which has an unknown composition that must be different from that of baryonic matter. The terms *baryonic* and *leptonic* come from Greek words meaning, respectively, "heavy" and "light.") Put another way, dark matter constitutes between 81 and 87 percent of all the matter in these galaxy clusters. Furthermore, galaxy clusters are thought to represent a fair sample of all the matter in the universe. The Chandra data thus imply that stars and gas constitute only about 16 percent of all the matter in the universe. The rest is dark matter.

✧

The evidence for dark matter is broad and deep and is not confined to X-ray observations of galaxy clusters. That work was preceded by optical and radio astronomers who had largely forgotten about Fritz Zwicky's work on clusters and were concentrating on the rotation rate of the outer regions of spiral galaxies.

Even though it takes a galaxy hundreds of millions of years to make one rotation, astronomers can determine how fast it is moving by using a few basic principles of light. The first is that light from a known type of atom gives off a characteristic pattern, or spectrum, of light. The second is that the wavelength of light from a moving source is shifted by an amount that depends on the motion of the source, an effect known as the Doppler shift, after Austrian physicist Christian Doppler (1803–53). If the source is moving toward you, shorter wavelengths are observed. If the source is moving away, the wavelength of the radiation is longer. The amount of the shift depends on the speed of the source. Traffic cops use the same principle to measure a vehicle's speed: They employ a radar gun to bounce a radar wave of known wavelength off a vehicle and measure the wavelength of the reflected wave to determine the vehicle's speed. By observing the spectrum of a portion of a galaxy and comparing it to a reference spectrum, the Doppler shifts of the light from the galaxy can be measured, and from that information, the speed of the gas producing the light can be determined.

In the 1970s, Vera Rubin and her colleagues used the Doppler technique to show that gas clouds on the edges of spiral galaxies are rotating so rapidly that the gravity of the observed stars and gas in the galaxies could not possibly hold onto the gas clouds (Rubin and Ford 1970; Ford, Rubin, and Roberts 1971; Rubin 1980; Rubin, Ford, and Thonnard 1980). A large amount of dark matter, with a mass five to ten times that of observed stars and gas, is required to explain these observations. Similar conclusions followed from observations made at about the same time by Morton Roberts and Arnold Rots, who used radio telescopes (Roberts and Rots 1973; Roberts 1976). Since then, supporting evidence has come both from observations of the smallest known galaxies and from an analysis of the cosmic microwave background radiation (figure 8), which is relic radiation due to the hot, primordial cosmic fireball associated with the origin of the universe almost 14 billion years ago.

Dwarf galaxies are faint, inconspicuous systems with only a few million stars, but ultimately they may play a key role in understanding dark matter. Measurements of the random motions of stars in nearby dwarf galaxies show that these galaxies may require a much larger fraction of dark matter than normal galaxies do. Since dwarf galaxies contain fewer sources of X-rays and gamma rays than normal galaxies do, any potential X-ray or gamma-ray signal produced by the decay or annihilation of dark matter particles should be easier to detect in dwarf galaxies.

Cosmic microwave background radiation reveals what the universe was like 13.7 billion years ago, when it was only a few hundred thousand years old, long before any galaxies and clusters of galaxies had formed. At this time, the universe was an expanding gas composed primarily of protons, electrons, photons, neutrinos, and dark matter. The intensity of cosmic microwave background radiation is very nearly—but not quite—the same in all directions. Small variations, expressed

8 | COSMIC MICROWAVE BACKGROUND RADIATION

A map of the cosmic microwave background radiation as detected with the greatest precision yet by the European Space Agency's Planck mission. The cosmic microwave background was imprinted onto the sky when the universe was about four hundred thousand years old. This map has been enhanced to show tiny temperature fluctuations. The temperature scale is in micro-Kelvins; a Kelvin represents the number of degrees Celsius above absolute zero (−273°C). The lowest-temperature areas (blue) are of enhanced density—the seeds of the stars and galaxies of today.

as a fraction of a percent, have been detected. These variations or fluctuations are due to clumps of matter that are either hotter or cooler than the average. The rate at which clumps grow in a hot, expanding gas can be calculated for different mixtures of photons, baryons, neutrinos, and dark matter, and the clumps leave an imprint on cosmic microwave background. A prominent imprint—called an acoustic peak—is left by sound waves created by collapsing clumps. Comparison of observations of cosmic microwave background by NASA's Wilkinson Microwave Anisotropy Probe (WMAP) and the European Space Agency's Planck satellite, based on the theory of

9 | BULLET CLUSTER

A composite image of the Bullet Cluster, which formed when two smaller clusters of galaxies collided. The pink clumps, which are hot gas detected by Chandra in X-rays, contain most of the subclusters' normal matter. The bullet-shaped clump at right is hot gas from one subcluster that passed through hot gas from the other, larger subcluster at left during their collision. The background optical image is from the Magellan Telescope and Hubble and depicts the galaxies in orange and white. The blue areas show the concentration of total mass, including dark matter, as determined by the gravitational lensing technique.

the acoustic peak, indicates that the universe contains about 85 percent dark matter and 15 percent baryonic matter. We can go back still farther in time, to the first few minutes of the Big Bang—not through direct observation, but by looking at the nuclear ashes from that fiery period. The critical nucleus here is deuterium, or heavy hydrogen, composed of one proton and one neutron. Ordinary hydrogen has only a single proton in its nucleus. Neutrons are slightly more massive than protons, and a free neutron decays into a proton in about fifteen minutes.

However, if the temperature is high enough, negatively charged electrons can be forced into positively charged protons to create neutrons. These conditions existed when the universe was less than three seconds old and the temperature was about 6 billion degrees Celsius. At that time, neutrons were about one-sixth as abundant as protons. The neutrons and protons could combine into deuterium, but gamma rays quickly destroyed the newly formed deuterium nuclei as long as the temperature remained above about a billion degrees Celsius, or until the universe was about two minutes old. After that time, deuterium nuclei could survive, and almost all the free neutrons combined with protons to form deuterium nuclei. Then, very quickly, in a matter of a couple of minutes, almost all the deuterium nuclei combined to form helium nuclei. By the time the universe was about twenty minutes old, it had cooled down and thinned out, and no more nuclear reactions would occur until atoms gathered into stars a few hundred million years later.

A few deuterium nuclei survived those first few minutes after the Big Bang and can be detected today in interstellar and intergalactic clouds. The amount that survived depends on the rate of the nuclear reactions that combined deuterium nuclei into helium nuclei. This rate, in turn, depends on how many deuterium nuclei were around to participate in the reaction, which depends on the density of matter in the universe at that time. Working backward, we can obtain an estimate of the fraction of baryonic matter in the universe that is fully consistent with the results of other methods: about 15 percent.

Another independent line of evidence points to the dominance of dark matter in galaxy clusters. According to Einstein's theory of general relativity, space is curved near strong gravitational fields. One consequence of the warping of space by gravity is that the path of light from background galaxies is bent when it passes near a cluster, in much the same way that a glass lens bends light. Images of galaxy clusters are distorted by this "gravitational lensing" effect by an amount that depends on the mass of the cluster. This method yields an estimate of the amount of dark matter in galaxy clusters that is in good agreement with estimates derived from X-ray observations.

Gravitational lensing, coupled with Chandra's X-ray observations, gives the most graphic and convincing evidence for the existence of dark matter. The accompanying image of the Bullet Cluster of galaxies (figure 9, page 20) shows hot, X-ray-producing gas (pink) and optical light from stars (orange and white) in the cluster galaxies. The X-ray observations show that the Bullet Cluster is composed of two large subclusters colliding at high speeds in one of the most energetic events ever observed. Using the gravitational lensing technique, astronomers have deduced that the total mass concentration in the clusters (blue) is separate from that of the hot gas. This separation was likely produced by the swift collision between the two subclusters. The hot gas in each subcluster was slowed by a drag force similar to air resistance. In contrast, the dark matter was not slowed by the collision because it interacts weakly, if at all, with itself or the gas except through gravity. The masses of dark matter passed through one another like ghosts and moved ahead of the hot gas, producing the separation seen in the image.

Although such violent collisions between clusters are rare, observations of a half-dozen other clusters, including MACSJ0025.4-1222 (figure 10), have confirmed the results from the Bullet Cluster. Together these findings make it very difficult to avoid the conclusion that most matter in the universe is dark matter.

However, not all scientists accept this conclusion. Since the first publication of evidence for dark matter in the 1980s, a group of skeptics has maintained that a modified law of gravity can explain all the observations purporting to show evidence of dark matter. Notable among these is Mordecai Milgrom of the Weizmann Institute in Israel, who abandoned a career as a conventional astrophysical theorist to pursue this path, publishing ninety-nine papers on his model in the past thirty-two years (see, for example, Milgrom 1983). In modified gravity theories, since there is no dark matter, the gravitational force field should be spatially coincident with that of the hot gas, contrary to what is observed in the Bullet Cluster. While Milgrom acknowledges that the Bullet and similar clusters pose a problem for his theory, he has not given up, saying in a recent publication that "just as everything that glitters is not gold, everything that is dark is not the dark matter. What is 'seen' in the Bullet might well be just an inkling of small amounts of yet undetected baryons indigenous to the cluster" (2015).

Until the nature of dark matter has been definitely established, we have to keep an open mind. In the meantime, most astrophysicists who work on such things assume that dark matter does exist and spend their time trying to find it or to work out how it influences baryonic matter, which we can see.

10 | MACSJ0025.4-1222 GALAXY CLUSTER

Image showing two subclusters of galaxies merging at high speeds. The distribution of ordinary matter, mostly in the form of hot gas, is shown in pink, and the distribution of total mass—mostly dark matter—is in blue. The galaxies are white and cyan. The distribution of hot gas (~70 million degrees Celsius), mapped by Chandra, is clearly displaced from the distribution of galaxies and dark matter. The image is 3.2 arc minutes across, corresponding to 4 million light-years at a distance of 5.6 billion light-years from Earth.

11 | EL GORDO GALAXY CLUSTER

A composite image of Chandra X-ray data (pink) overlaid on Hubble optical data (red, green, and blue) from the galaxy cluster ACT-CL J0102-4915, 7 billion light-years from Earth. Astronomers dubbed this cluster El Gordo ("the fat one" in Spanish) because of its tremendous mass, which is estimated to be that of 3 quadrillion Suns. The distribution of its total mass (shown as the blue clouds)—which is about 80 percent dark matter—was determined by gravitational lensing. The X-ray image gives El Gordo a cometary appearance. Alongside the optical data, this image reveals that El Gordo is the product of two galaxy clusters colliding at several million miles per hour. As in the Bullet Cluster (figure 9), the hot gas in each cluster was slowed down by the impact, but the dark matter was not. This image captures a region about 7.7 million light-years across.

An attractive feature of this cold dark
matter hypothesis is its considerable
predictive power.

George R. Blumenthal et al., "Formation of Galaxies and Large-Scale Structure
with Cold Dark Matter," 1984

COLD DARK
MATTER

Given the view that the evidence for dark matter is compelling, one of the most pressing questions in astrophysics, indeed in all of physics, is: What *is* it?

We do not have an answer to this question, but it is becoming increasingly clear what it is not. As discussed in the previous chapter, observations of the amount of deuterium in the universe, the detailed structure of fluctuations in the cosmic microwave background radiation, and Chandra's observations of galaxy clusters all agree that there is about five to six times more dark matter than baryonic matter in the universe.

Black holes would seem to be the ideal dark matter candidate. But stellar mass black holes, produced by the collapse of massive stars, are made of baryonic matter, so they won't work. However, it is possible that primordial black holes were created in the very early universe, before it was a few seconds old. In that event, primordial black holes already would have been in place by the time the nuclear reactions creating deuterium and helium occurred, and they would not have participated in the processes that set the scale of the so-called acoustic peak in the cosmic microwave background.

In 1974, Stephen Hawking of Cambridge University in the United Kingdom published a paper in which he predicted that black holes would evaporate in a few billion years by a process that came to be known as Hawking radiation. This radiation occurs because empty space, or the vacuum, is not really empty. Instead it is a seething sea of virtual particles of every type that pop into and out of existence for an infinitesimally brief time. This is possible because, according to the uncertainty principle of quantum theory, energy may be borrowed from the vacuum, but it also must be repaid quickly. The greater the amount borrowed, the quicker it must be repaid. For example, the energy to make a pair of virtual particles consisting of an electron and a positron—the electron's antiparticle—must be repaid in one sextillionth of a second or, to use the official scientific term, one zeptosecond.

Hawking showed that if a pair of virtual particles is created near a black hole, there is a chance that one of them will be sucked into the black hole before it is destroyed, while its partner escapes into space. In this event, a virtual particle will have become real. The energy for this came from the black hole, so the black hole slowly loses energy and mass by this process.

Hawking radiation is predicted to be weak and has never been detected, but a few skeptics notwithstanding, most experts in black hole astrophysics think that some version of this process must occur. One consequence of Hawking radiation is that black holes with masses much less than 1 trillion kilograms (about the mass of a small comet) would have radiated all their mass away in the 13.8 billion years since the Big Bang. Black holes with masses larger than comets can be detected in principle by gravitational microlensing—a detectable blip would show up in the radiation from a background star as the primordial black hole moved in front of the star. So far these searches have come up empty. For now, few astrophysicists take the primordial black hole option for dark matter seriously and have focused their attention instead on an as-yet unidentified type of particle.

The fact that no known type of particle fits the requirements for dark matter has profound implications for physics as well as astronomy. The so-called Standard Model of particle physics, according to which all atoms are composed of nuclei and electrons, and all nuclei are composed of just six types of quarks, has had many successes in explaining the existing types of elementary particles and predicting the existence of once-undiscovered particles. However, it must be revised to account for dark matter.

Among particle physicists, a favorite category of candidates involves super-symmetry, a proposed extension of the Standard Model. Known particles fall into one of two general classes: they are either fermions, named after the Italian physicist Enrico Fermi, or bosons, named after the Indian physicist Satyendra Bose. The protons, neutrons, and electrons that make up all the elements are loners—they don't like to share the same quantum state with other fermions. This is a good thing; otherwise matter as we know it—including ourselves—wouldn't exist. The electrons in each atom must occupy successively larger shells, or orbits, so atoms under normal conditions found on Earth cannot be squeezed too closely together, which is why everything isn't a dense blob. Bosons, in contrast to fermions, are more gregarious, or less discriminating, depending on your point of view. Many bosons can occupy the same state. The most familiar boson is the photon, the particle of light. And, of course, there is the Higgs boson, a crucial particle in the Standard Model. The Higgs boson was named after Peter Higgs, a British physicist and one of six physicists who proposed the existence of such a particle in 1964. The Higgs boson is postulated to originate from an invisible field, the Higgs field. This field fills up all space, and it plays a fundamental role in determining how baryons and leptons acquire mass. The Higgs was discovered in 2012 by physicists working at the Large Hadron Collider (LHC) in Geneva, Switzerland. The following year, Higgs and Francois Englert, a Belgian physicist, shared the Nobel Prize in physics for their work on the Higgs boson.

The general class of theories called supersymmetry would link each fermion to a boson partner and vice versa. A photon would have a fermionic partner, the photino; an electron, a bosonic partner, the selectron; and so on. These particles are often lumped under the general name of neutralino, since they must be electrically neutral; otherwise they would interact with the electromagnetic field and produce light. According to supersymmetric theories, neutralinos would have been produced in abundance in the extremely high temperatures of the Big Bang's early moments, and enough would have survived to account for dark matter.

It all comes down to bookkeeping, or balancing accounts. One of the many profound implications of Einstein's famous equation $E = mc^2$ is that a collision of two particles, say photons, with sufficiently high energies can create new pairs of particles and their antiparticles. At the high temperatures and densities of the early moments of the Big Bang, the universe was filled with extremely high-energy photons, and they created an enormous number of particles and antiparticles—antiparticles of a given type of particle have the same mass as their corresponding particle, but an opposite charge. Some of these particle-antiparticle pairs may have been dark matter particles. The reverse process to the creation of particle-antiparticle pairs occurs with particles and their antiparticles annihilating each other to produce two photons. The rate of annihilation depends on the rate of collisions, which in turn depends on the density of particles and their speeds, both of which are very large early on but decrease rapidly as the universe expands and cools. The primordial particle soup contained a rich variety of particles whose abundances were determined by the equilibrium between creation and destruction. The situation was like a cosmic bank account, with deposits and withdrawals going on continually, the balance in any one particle's account being determined by the rates of deposits and withdrawals.

The ongoing cooling effectively closed out the accounts of various particles depending on their energies. High-energy accounts, corresponding to more massive particles, were closed out before low-energy ones, because it takes more energetic photons to create more massive particles. In the early stages of the expansion, this had the effect of draining the accounts of the massive particles—no new ones were created, and the old ones were annihilated with their antiparticle partners.

At a critical point early in its evolution, the universe was still hot enough for particle-antiparticle pairs to be created by collisions between high-energy photons, but the gas expanded and thinned out before the particles could find each other to annihilate. A particle's antiparticle was out there somewhere, but the chance of the two meeting was quickly dwindling. The particles then "froze out," existing with their number determined by the details of the expansion and the masses of the particles.

Using the bank account analogy, the assets—that is, the abundances of a particular particle-antiparticle pair—are frozen. The freeze-out process implies that a weakly interacting massive particle, or WIMP, could have been created in the first trillionth of a second of the Big Bang. The mass of the hypothetical WIMP is very uncertain, but most estimates place it somewhere in the range of a few to several hundred times the mass of the proton.

WIMPs are the favorite dark matter candidates, but other possibilities are under active consideration, including sterile neutrinos, axions, asymmetric dark matter, mirror dark matters, and extradimensional dark matter. Of these, sterile neutrinos have attracted the most attention recently because of the possibility of observing their decay into X-rays (see chapter 4).

While particle physicists struggle to come up with a modification of the Standard Model that can account for dark matter, using the LHC to create conditions similar to those in the first trillionth of a second of the Big Bang, astrophysicists look out into space and into simulations produced by supercomputers for guidance.

Astrophysicists don't generally take sides in the debate over the exact nature of dark matter. Rather, they group dark matter particle candidates into three broad categories related to the speed of the hypothetical particles at the crucial time when the clumps of gas that would form galaxies and clusters of galaxies first began to grow:

Hot dark matter: Particles such as the known types of neutrinos, which were moving at near the speed of light when the clumps that would form galaxies and clusters of galaxies first began to grow

Cold dark matter: Particles that were moving slowly when the pregalactic clumps began to form

Warm dark matter: Particles with speeds between those of hot and cold dark matter

Hot dark matter particles would have been moving so rapidly that clumps with the mass of a galaxy would have quickly dispersed. Only clouds with the mass of thousands of galaxies—that is, those the size of galaxy clusters—could have formed. These clouds would have then fragmented into smaller clouds that then fragmented into the still smaller clouds that would become galaxies. So, looking back in time, one would expect to find clouds with no galaxies; a little later, smaller clouds; and then clouds fragmenting into a group of galaxies in the process of forming. Something similar is observed in the process of star formation. Large clouds fragment into smaller clouds, which fragment into groups of stars. It seems reasonable to expect that the process might work that way on a larger scale, with galaxies.

However, the universe did not cooperate with this reasonable plan. It seemed to favor the bottom-up process wherein galaxies form first. This can be achieved with cold dark matter, which can form into clumps of galaxy-sized mass or smaller. Galaxies would form first; then galaxies would merge into groups; and groups would gather into clusters in this reversed process. The cold dark matter model led to a major breakthrough in understanding how galaxies form. Without it, astrophysical model builders were stuck. To be sure, clumps would form under the influence of gravity in the primordial gas of the Big Bang, but the clumps would not grow quickly enough to form galaxies.

For the first few hundred thousand years, normal matter consisted mostly of protons and electrons roaming freely in the primordial gas. It was still too hot for electrons to combine with protons to form hydrogen atoms. Frequent collisions between photons and electrons had the effect of creating pressure, which prevented clumps of matter from collapsing too quickly. Essentially the same thing happens when you compress the plunger on a tire pump.

Once the electrons and protons had combined to form atoms, the number of collisions between photons and electrons dropped drastically, leading to a drop in pressure, so that the force of gravity could compress the clumps. However, computer simulations indicate that this was too little, too late. It was difficult to grow galaxies quickly enough to explain observations that galaxies of billions of stars had already formed a few billion years after the Big Bang. A few billion years is a long time, but there was the competing factor of the expansion of the universe. If gravity couldn't pull the gas together quickly, it just might not happen at all, the models suggested, so we would be stuck with a universe with no galaxies, no stars, and no planets. Clearly, this was not acceptable.

Cold dark matter came to the rescue. Because dark matter is not affected by photons, clumps of dark matter do not feel photon pressure and can collapse and pull in more dark matter, thereby jumpstarting the formation of galaxies. This breakthrough, concurrent with advances in numerical methodologies and computing speed, has allowed, in the words of Rachel Somerville and Romeel Davé (2015), "extraordinary progress in our ability to simulate the formulation of structure within the paradigm of the Cold Dark Matter model."

Detailed computer simulations of the formation of galaxies and clusters of galaxies have supported these general arguments. The Chandra images provide many examples of clusters being constructed by the merger of groups and subclusters of galaxies (figure 11, page 24, and figure 12). There seems to be little room for doubt that the basic outline of the cold dark matter hypothesis is correct on a large scale. However, as Somerville and Davé cautioned, "It is almost surely the case that the physical processes included in models so far will not be sufficient to fully describe galaxy evolution, and there will be many twists and surprises forthcoming" (ibid.).

One of the appealing aspects of dark matter is its simplicity. Whatever it may be, it just *is*. That makes it easy to deal with in computer simulations that seek to explain the formation of galaxies and clusters of galaxies. It doesn't interact with other matter except through its gravity, so it isn't whipped around by magnetic fields; nor does it lose energy by radiating; nor does it bounce around in collisions with other particles. Or does it? As computer models and observations have gotten more sophisticated, some problems with the cold dark matter paradigm have surfaced.

On a smaller scale, in the central regions of galaxies and in dwarf galaxies, there are indications that some tweaks to the model may be needed. The problem is that the collapse of a galaxy under the influence of the gravity of cold dark matter should produce a sharp spike, or cusp, in the concentration of dark matter in the center of the galaxy. But astronomers have not observed such a cusp. Instead, the dark matter concentration increases smoothly toward the center. This problem was noticed early on, but researchers generally assumed—or hoped—that lack of a dark matter cusp was an artifact of the limitations of both the coarse-grained nature of the computer simulations and the difficulty of measuring the amount of dark matter in the central regions of galaxies and clusters of galaxies. Even after two decades of dramatic improvements on both fronts, the discrepancy remains. Optical observations of the central regions of galaxies have confirmed it. A team of astronomers led by Andrew Newman of Carnegie Observatories used Chandra X-ray data on the cluster Abell 383, together with optical data from Hubble and the Subaru and Keck telescopes on Mauna Kea in Hawaii, to show that the concentration of dark matter is leveling off, rather than peaking sharply, in the central regions of this cluster.

Recent simulations suggest that taking into account the injection of energy into central regions by supernovas or activity associated with the presence of a supermassive black hole, or the effects of small galaxies falling into larger ones, may be able to solve part, if not all, of the problem on a galactic scale. However, it is not clear that any of these processes work for galaxy clusters. If the relative lack of dark matter in the center of Abell 383 is confirmed, it may be evidence that dark matter particles can interact with one another through some unknown, nongravitational force, and a revision in the cold dark matter model may be in order.

Collisions between galaxy clusters provide a laboratory for measuring the interaction of dark matter particles. The chance that the stars in colliding clusters will run into one another is extremely small. Even though the stars are large, they are very far apart and glide past one another like ships passing in the night. In contrast, the clouds of hot, X-ray-emitting gas are huge, and as two clusters collide, the effect is much like that of the wind experienced by a speeding motorcyclist. If the motorcyclist's helmet is not on securely, the wind, or ram pressure, created by his motion will blow the helmet away. In the same way, when two galaxy clusters collide at speeds of several million miles per hour, the gas clouds experience ram pressure and are slowed down, while the stars continue on their way.

Dark matter, if it collides with itself, should also lag behind the stars, though not as much as the hot gas does. In the Bullet Cluster, the lack of any observed separation of dark matter from the stars sets a limit on the frequency of collisions between two dark matter particles. This limit allows for the possibility that collisions between dark matter particles can explain the lack of cusps in the dark matter concentration in the centers of galaxy clusters. A study led by David Harvey of the École Polytechnique Fédérale de Lausanne in Switzerland showed that Chandra and Hubble Space Telescope observations of seventy-two collisions between galaxy clusters imply that the cross-section (an estimate of the probability of a collision between two particles) of dark matter self-interactions is less than half of previous limits. This limit comes very close to shutting the door on the collisional dark matter option. It seems likely that some other process causes the observed lack of a cusp in the concentration of dark matter in galaxy clusters' central regions. More observations are needed to confirm the absence of a cusp, however, along with more detailed simulations that take account of energy injected into central regions, or the stirring up of dark matter by smaller, satellite galaxies falling into central regions of galaxy clusters.

12 | EMSS 1358+6245 GALAXY CLUSTERS

A Chandra image of these clusters, about 4 billion light-years away in the constellation Draco. When combined with Chandra's X-ray spectrum, this image allowed scientists to determine that the mass of dark matter in the clusters is about four times the mass of normal matter. The relative percentage of dark matter increases toward the clusters' center, and measuring the exact amount of this increase enabled astronomers to understand limits on the rate at which the clusters' dark matter particles collide with one another—information extremely important to the quest to comprehend the nature of dark matter, which is thought to be the most common form of matter in the universe.

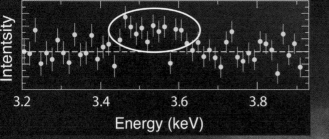

13 | PERSEUS CLUSTER

A Chandra image of hot gas in the central region of the Perseus Cluster.
Low-energy, medium-energy, and high-energy X-rays are color-coded red, green,
and blue, respectively. The area shown is about 768,000 light-years across.
A potential dark matter decay signal is represented by the circled data points in
the inset, which is a plot of X-ray intensity as a function of X-ray energy (keV).

4

Come, Mister tally man, tally me banana

Daylight come and me wan' go home.

Traditional Jamaican folk song

DARK MATTER
GOING BANANAS

A ubiquitous feature of modern life is the barcode. Since they were introduced more than fifty years ago, barcodes have been used to identify people, products, and locations. Hardly a day passes when we do not encounter one. Likewise, hardly a day passes in the work of an astrophysicist when he or she does not look at, or read a reference to, the cosmic barcodes called spectra.

Astronomers and physicists began using spectrographs more than two centuries earlier, after Joseph Fraunhofer, a German glassmaker and optician, invented the first spectroscope and used it to identify spectral lines from fire and the Sun. In the 1860s, the German physicist Gustav Kirchhoff and chemist Robert Bunsen undertook an experimental program to examine and classify the spectra of various chemical compounds. They showed that spectroscopy can be used to identify the chemical elements present in a compound and that some of these elements are present in the Sun. About the same time, the English astronomers William and Margaret Huggins used a spectrograph to show that stars are composed of the same elements that are found on Earth. This discovery was of fundamental importance. Along with Newton's universal law of gravitation, it implied that the physical laws that govern the structure of atoms—the structure that governs the spectra from the elements—are the same throughout the universe.

The origin of spectral lines was not understood until the development of quantum theory in the early years of the twentieth century. According to quantum theory, electrons bound to a nucleus to form an atom are not free to move in an arbitrary manner. Their motions are strictly regulated by the quantum rules that govern the structure of an atom. These rules restrict the electrons' motion to specific energy levels, or quantum states. These levels can be thought of as stair steps. If you wish to move up or down the stairs, you must move from one step to another one. You cannot move to a position between the steps. You might move more than one step at a time, but you cannot take half steps. Likewise, an electron in an atom can move only from one level or quantum state to another one, and even then, some moves are much likelier than others. The likelihood, or probability, of a given move can be calculated from the principles of quantum mechanics.

As electrons jump from one quantum state to a lower one, they give off photons. The energy of these photons is related to the magnitude of the quantum leap. Because of the quantum restriction that electrons must jump from one distinct level to the next, the photons must have distinct energies determined by the arrangement of the atomic levels.

Radiation from a hydrogen atom, for example, is not spread smoothly over a rainbow of colors from red to blue. Rather, it is concentrated in a few extremely narrow-energy—or, equivalently, wavelength—bands called spectral lines. A careful study of the pattern of radiation given off by hydrogen atoms reveals a barcode for the levels of the hydrogen atom. The atomic barcode for all hydrogen atoms is the same. A hydrogen atom extracted from the water in the birdbath in your backyard has the same basic structure as one on Jupiter or in a cloud of interstellar gas. Hydrogen is hydrogen, wherever you find it—and you can find it by looking for a set of photons that match the barcode.

In this way, astrophysicists have determined that stars are mostly made of hydrogen, with a mixture of helium and traces of heavier elements such as carbon, nitrogen, oxygen, iron, and so on, which are present in concentrations similar to that of the Sun—that is, a level of a few percent. Roughly the same proportion holds true for gas between the stars and in galaxy clusters, although there are significant variations, and these differences are crucial clues to the history of the gas. Heavier elements are created in stars and then ejected into space both gradually and explosively, causing variations in the quantity of heavy elements in the interstellar gas.

The oldest known stars formed more than 13 billion years ago from relatively pristine gas. Observations show that the abundances of heavy elements in these stars are thousands of times lower than those in the Sun, a clear indication that the interstellar gas from which new stars form is enriched gradually over the eons. Intergalactic gas in galaxy clusters exhibits an abundance of heavy elements, about 30 percent that of the Sun. While not nearly as extreme as the deficit in the oldest stars, this difference does indicate that the buildup of heavy elements, likely due to ejection of matter by supernovas in galaxy clusters, is an ongoing process. The study of X-ray spectra of galaxy clusters is one primary means of studying this enrichment, which has implications for the frequency of supernovas in the galaxies in a cluster and the type of supernovas primarily responsible for the enrichment.

In 2001, Kevoork Abazajian of the University of California, Irvine, and his colleagues suggested that the X-ray spectra of galaxy clusters might also contain a weak, though detectable, signature of dark matter. Their idea was based on a proposal that the dark matter is in the form of sterile neutrinos, hypothetical particles that have been proposed as a candidate for dark matter. Sterile neutrinos, if they exist, would decay over a period of billions of years into X-rays.

A Chandra study of the central region of the Perseus galaxy cluster, shown in figure 13, page 34, has revealed an unidentified X-ray signal in the data. This signal is represented by the circled data points in the inset, which is a plot of X-ray intensity as a function of X-ray energy (expressed in kiloelectron volts, or keV). The signal is also seen in data from seventy-three other galaxy clusters, gathered using the European Space Agency's XMM-Newton observatory. This unidentified X-ray emission line—a spike of intensity at a very specific energy, in this case centered on about 3.56 keV—has generated considerable excitement and controversy.

One intriguing possible explanation for this X-ray emission line is that it is produced by the decay of sterile neutrinos. The detection of the emission line is pushing the capabilities of both Chandra and XMM-Newton in terms of sensitivity, however, so the scientists involved have been cautious. "We know that the dark matter explanation is a long shot, but the pay-off would be huge if we're right," said Esra Bulbul (2014c) of the Harvard-Smithsonian Center for Astrophysics, who led the study. "So we're going to keep testing this interpretation and see where it takes us." Coauthor Maxim Markevitch (2014), at NASA's Goddard Space Flight Center, agreed. "We have a lot of work to do before we can claim, with any confidence, that we've found sterile neutrinos. But just the possibility of finding them has us very excited."

It also has many other scientists excited. Within months after Bulbul and her colleagues had placed their paper on the Astro-ph archive, a publicly accessible database of astrophysics papers, it ignited a flurry of activity. The work was cited in fifty-five new papers, which mostly involve theories discussing the emission line as possible evidence of dark matter. Some papers explore the sterile neutrino interpretation, but others suggest that different types of candidate dark matter particles, such as the axion, may have been detected.

A different group, led by Alexey Boyarsky of Leiden University in the Netherlands, placed a paper on Astro-ph just ten days after Bulbul et al. had done so, reporting evidence of an emission line at the same energy in XMM-Newton observations of the galaxy M31 and the outskirts of the Perseus cluster. This paper strengthened the evidence that the emission line is real and not an instrumental artifact.

However, some scientists were excited in a different way. Soon papers expressing deep skepticism of the discoveries were posted on Astro-ph. The attacks were on two fronts. One argued that the data had not been analyzed properly and that the signal simply doesn't exist. Another held that the emission line can be explained in more prosaic terms by highly ionized potassium, chlorine, or argon atoms. The possibility that potassium could explain the result led Tesla Jeltema and Stefano Profumo of the University of California, Santa Cruz, to title an Astro-ph paper "Dark Matter Searches Going Bananas: The Contribution of Potassium (and Chlorine) to the 3.5 keV Line" (2014a), in reference to the high potassium content of bananas. Journal editors apparently did not appreciate the attempt at humor, and the title was changed before publication to the more prosaic "Discovery of a 3.5 keV Line in the Galactic Centre and a Critical Look at the Origin of the Line across Astronomical Targets" (2015).

Bulbul and her colleagues had acknowledged that potassium and argon, but not chlorine, could possibly explain the mystery line, but they argued that the amount of those elements would have to be ten times what was normally observed in other places in the cosmos. Too many bananas would be required, in other words, although they refrained from saying that. Not necessarily, countered Kenneth Phillips of the Natural History Museum in London and coauthors Barbara Sylwester and Janusz Sylwester of the Polish Academy of Sciences in Wroclaw. Greatly enhanced abundances of potassium ions have been observed in flares on the Sun and other stars. Finally, Michael Anderson of the Max Planck

Institute for Astrophysics in Munich and colleagues searched for the mystery line in 170 galaxies with Chandra and XMM-Newton and came up empty (Anderson, Churazov, and Bregman 2015). The lack of detection may not be a problem, some theorists responded, saying that some channels for producing the neutrino decay occur preferentially for fast-moving dark matter particles, which are more likely to exist in massive galaxy clusters. Another possibility is that the conversion of sterile neutrinos into X-rays requires a magnetic field of a certain strength, so it is not surprising that it is observed in some objects and not others. The only conclusion that can be drawn from these and dozens of other papers is that at this point there is no answer as to whether sterile neutrinos can explain dark matter.

Sterile neutrinos are the latest dark matter candidate to attract attention and generate food fights among astrophysicists, but they are by no means the only ones. Chief among the other points of contention are cold dark matter WIMPs (weakly interacting massive particles), discussed earlier, which are expected to decay into gamma rays and quarks; indeed, more than a dozen groups have claimed to have detected a gamma-ray signal of dark matter decay. However, as Tansu Daylan of Harvard University and his colleagues have commented: "Most, if not all, of these signals, have nothing to do with dark matter, but instead result from some combination of astrophysical, environmental, and instrumental backgrounds. . . . Given the frequency of such false alarms, we would be wise to apply a very high standard before concluding that any new signal is, in fact, the result of annihilating dark matter." Daylan et al. then went on to say that this criticism does not apply to their research: "There are significant reasons to conclude, however, that the gamma-ray signal described in this paper is far more likely to be a detection of dark matter than any of the previously reported anomalies." These reasons include a strong signal and a simple explanation for it, namely the decay into quarks "without any baroque or unexpected features" (2014).

However, Profumo, again playing the role of skeptic, suggested along with coauthor Eric Carlson that Daylan and colleagues' result is just one more false alarm and that what is "baroque" may be in the eye of the beholder. In a recent Astro-ph posting, they argued that the Galactic Center excess is readily explained by a recent injection of high-energy protons and concluded that "it is premature to argue that there are no standard astrophysical mechanisms that can explain the excess" (2014).

There is thus still ample room for doubt that the mystery of dark matter has been solved, and the search goes on not just in the sky, but also on Earth, and in some extraordinary places, such as the Large Hadron Collider. This work has little to do with Chandra, but it could have a lot to do with the search for dark matter.

If WIMPs (weakly interacting massive particles) are the dark matter, then we must be swimming in a sea of dark matter, and a billion or more WIMPs must pass through our bodies every second. So why do we need a telescope to find them? Why not build a detector here on Earth? Many physicists have thought the same thing, and some have spent a good part of their careers raising money to create instruments to detect dark matter, building them, searching in vain for their target, and then raising more money to build even larger detectors.

The problem is the "weakly interacting" part of a WIMP. The experiments that have searched for dark matter have been of a heroic nature. The Large Underground Xenon (LUX) detector contains 368 kilograms of liquefied xenon in a canister about the size of a phone booth. The detector sits in a twenty-five-foot tank of ultra-pure water about a mile deep in an abandoned gold mine in South Dakota. The DAMA/LIBRA experiment, which uses a few hundred pounds of sodium iodide crystals, is buried about a mile underneath the Gran Sasso, a mountain in Italy. The basic goal of these experiments is to detect an extremely faint flash of light produced by the interaction of a WIMP with the detector.

So far the LUX experiment has come up empty, but scientists in the LUX collaboration continued another run lasting until the end of 2016. However, other scientists think that LUX is simply not big enough, that detectors with one hundred times more mass will be needed to see WIMPs at the rates expected. A planned upgrade, scheduled for 2019, will use about twenty times as much liquid xenon as the present version.

In contrast to the report of a negative result from the LUX experiment, the DAMA/LIBRA experimenters have claimed a positive detection. However, few other scientists are cheering. The consensus seems to be that the signal is real but is likely produced by something other than dark matter, such as neutrons leaching out of rocks and decaying to produce the observed signal. As with the controversy over sterile neutrinos, this skepticism has strained relations among erstwhile friends in the physics community.

Rather than take the passive approach of observing dark matter directly in the lab or indirectly through astronomical observations, particle physicists propose making the stuff. Since dark matter particles presumably were created in the first few nanoseconds of the Big Bang, when temperatures were a quadrillion degrees Celsius, a particle accelerator that reproduces these conditions might create dark matter.

The LHC may be the machine to do this. In a six-mile-circumference tunnel underneath Switzerland and France, this amazing machine has accelerated protons to 99.999994 percent of the speed of light and crashed them head-on into each other. In 2012, scientists found convincing evidence in the debris for the long-sought Higgs boson. In the next run, which is just beginning, the speed will be increased to 99.999998 percent of the speed of light. That doesn't seem like much of an increase, but at these speeds, this incremental increase will double the energy of the particles. This research is entering a regime of energy where physicists hope to see some interesting things, including evidence for dark matter, maybe in the form of squarks or sneutrinos, the hypothesized supersymmetric partners to quarks and neutrinos.

Some theoretical ideas suggest that dark matter particles have masses and energies beyond the reach of the LHC. The Cherenkov Telescope Array—an international project that will build more than one hundred ground-based telescopes in the next decade to capture light flashes from gamma rays scattered by the atmosphere—will open the window to energies ten times that of the LHC.

For the pessimists, there are what Mario Livio and Joe Silk have termed worst-case scenarios: "First, dark matter may not comprise one type of particle—as many current searches assume—but many. Second, the particles might interact only gravitationally, and could be practically invisible to conventional detectors" (2014).

Then there is what many consider the worse-than-worst-case scenario: Dark matter doesn't exist. Livio and Silk reflected the attitude of most astrophysicists when they said, "Most researchers think that we are far from needing new physical laws, especially because experimental avenues are still open. But unpleasant surprises are always possible" (ibid.). What constitutes an unpleasant surprise, however, depends on who is surprised.

14 | ABELL 2052

A composite image of the Abell 2052 galaxy cluster, about 450 million light-years from Earth. Chandra X-ray data are shown in blue, and optical data from the European Southern Observatory's Very Large Telescope are in orange. Abell 2052 is one of the eighty-six galaxy clusters that researchers have used to show that galaxy clusters' observed rate of formation and growth over the past 7 billion light-years has slowed, providing evidence of an accelerating universe. The huge spiral structure of multimillion-degree gas detected by Chandra spans almost a million light-years and reveals one effect of cluster growth: a small cluster of galaxies fell into a larger one and caused the hot gas to slosh around under the combined gravitational pull of the two clusters. The image is about 1.3 million light-years across.

5

Deep into that darkness peering,

long I stood there wondering, fearing.

Edgar Allan Poe, "The Raven," 1845

THE WONDERFUL— AND FEARFUL— DARK SIDE

P eering deep and long into the darkness is what astronomers do. What they have found there in recent years certainly inspires a sense of wonder while also attesting to major unseen features of the universe—dark stars, dark matter, and dark energy.

About four hundred years ago, the great astronomer Johannes Kepler realized that the very fact that the sky is dark at night reveals something profound about the nature of the universe. If the universe were infinite and uniformly populated with luminous stars, Kepler reasoned, then any line of sight must eventually encounter the surface of a star. Therefore, no dark gaps should lie between the stars, and the sky should not be dark at night. Kepler's solution was that the universe is finite rather than infinite. However, three hundred years later, Albert Einstein's theory of gravity showed that a finite universe must be a curved one in which light does not escape, so Kepler's solution did not work.

Absorption of distant light by interstellar gas does not solve the question either, because the gas would eventually heat up and glow. Clustering the stars and galaxies with progressively larger gaps between them could work in principle, but this would violate observations of galaxies and cosmic background radiation, which show that the universe is uniform on the largest scales.

The solution, which Edgar Allan Poe hinted at more than a century and a half ago in his prose poem "Eureka," is that the universe is not static; neither is it infinitely old. An overwhelming body of evidence supports the Big Bang theory, which postulates that the universe as we know it has been expanding and evolving from an extremely dense state that existed about 14 billion years ago. The stars do not live forever. Their lifetime is too short for light from all the stars in the universe to travel to Earth and fill the night sky uniformly with starlight. Most stars have faded out, leaving many dark, blank spaces in the night sky. Add to this the observation that the expansion of the universe is shifting the light to longer wavelengths that are invisible to human eyes, and we get a dark night sky.

In the past few years, astronomers have observed that the sky may be even darker than we thought. Light detected from a particular type of stellar explosion, called a Type Ia supernova, in galaxies billions of light-years away is fainter than we would expect for the estimated distance. There are three possible explanations: light from such supernovas has been absorbed by dust or atomic particles on its way here; the supernovas were not as bright in the distant past; or the estimated distance is wrong, so the supernovas are farther away than thought and thus appear fainter.

Dimming due to dust along the way can be ruled out. In previous years, a lot of work had gone in to showing that the intrinsic brightness, or luminosity, of Type Ia supernovas is well understood. This work was based on studies of many relatively nearby explosions of this type. It is possible that the type of supernova that can be detected from a distance of billions of light-years is different, or that stars exploded differently back then, but a careful study of hundreds of supernovas shows that this explanation is unlikely. The best explanation for the faintness of distant supernovas is that the galaxies are more distant, by about 10 to 15 percent, than originally thought. This explanation leads to the astonishing conclusion that the expansion of the universe is not slowing down, as would seem intuitively obvious—the gravity of the galaxies should be pulling back on the expansion and slowing it down. Rather, the expansion is speeding up. It is as if you threw a rock into the air, and instead of decelerating and falling back to Earth, it accelerated away from Earth into space.

How could astronomers have missed such a fundamental fact about the universe? After all, the expansion of the universe was discovered almost a century ago.

How difficult is it to measure the distance to a galaxy if we have the Hubble Space Telescope, as well as the other powerful telescopes on mountaintops in Chile and Hawaii, at our disposal? How could astronomers get the distances so wrong as to change our ideas about the makeup of the entire universe?

The short answer: Determining distances to galaxies is harder than you might think. Much harder. Edwin Hubble, namesake of Hubble's law, the Hubble constant, and the Hubble Space Telescope, expressed the challenge eloquently: "Eventually we reach the utmost limits of our telescopes. There we measure shadows and search among ghostly errors of measurement for landmarks that are scarcely more substantial" (1936).

The story of the expanding universe begins, as do many such stories, with the work of Albert Einstein. In 1905, Einstein published his special theory of relativity, which shows that our notions of space and time are relative, in that they depend on our motion through space and time. Einstein showed that the exact nature of this dependence can be determined by combining two principles. The first is the principle of relativity, which states that experiments with natural phenomena yield the same result in two different laboratories, even if those laboratories are moving in relation to each other, as long as the motion is uniform. For example, if you hold a ball at arm's length and drop it, it will land an arm's length from your feet, regardless of whether you are standing still or moving uniformly on a train or airplane. The second principle is the constancy of the speed of light in a vacuum. In other words, a measurement of the speed of a pulse of light would be the same regardless of whether the measurement was made on a moving train or in the railroad station.

The theory was a stunning success, explaining many puzzling aspects of the behavior of light. It's all relative, Einstein showed, meaning that whereas relative motion does not affect the basic underlying laws of physics, it does affect space and time in a predictable manner. For example, the measured lifetime for the decay of a radioactive particle is different for two observers, if one observer is moving relative to the other. Einstein further showed that relative motion also affects the measured energy of a particle in a predictable way, a conclusion that leads to the most famous equation in physics: $E = mc^2$, where c is the speed of light and E is the energy of the particle of mass, m. An interesting corollary of this equation is that photons, the massless particles of light, can be thought of as having a mass equivalent to their energy. The implication, according to Einstein, is that photons are affected by the gravity of stars. For example, light from a distant star could be deflected by the gravitational field of a foreground star. Indeed, the light emitted by a star is also affected by its own gravity.

An understanding of the effects of gravity on light requires a deeper understanding of gravity, a problem Einstein attacked next. He called the theory he published in 1905 the special theory of relativity because it addresses the special case of relative motion at a constant speed. Being Einstein, he was not content to leave it at that, and he started to investigate what happens when relative motion is not necessarily uniform. In other words, he wanted to apply his theory not just to the special case of uniform motion, but also to the general case of relative motions that involve acceleration. A major, omnipresent source of acceleration is acceleration due to gravity, so he was, in essence, working on a theory of gravity. After about a decade of very hard work, Einstein published his general theory of relativity.

In Newton's theory of gravity, the presence of a mass creates a gravitational force field, described by Newton's law of gravity, and this force field is plugged into Newton's laws of motion to calculate the motions. This recipe does an excellent job of calculating the motion of baseballs, satellites, and the like in Earth's gravitational field; planets and comets in orbit around the Sun; and the motion of stars in their galaxies and of galaxies in their clusters. Excellent, but not perfect. In certain extreme cases, it produces the wrong result. This is because Newton's laws ignore the effects of gravity on space and time.

When Einstein generalized his theory of relativity to include these effects, he came up with two equations analogous to Newton's. The first describes the force field but in a different way. It describes how space-time is curved in a specified way by gravity. The second set of equations, equivalent to Newton's laws of motion, describes how light and matter travel through this curved space. For example, light traversing space does not travel in a straight line but along a curved path. In essence, a foreground star acts as a gravitational lens and deflects light from a distant background star. In a dramatic and widely publicized test, this prediction was confirmed in a solar eclipse in 1919. The light passing by the Sun from a distant star was observed to be deflected by the amount predicted by Einstein.

As telescopes have become more powerful, gravitational lensing has become a standard tool used by astronomers to measure the masses of distant galaxies and clusters of galaxies, as discussed earlier in the context of measuring the amount of dark matter in the Bullet and other clusters. A much more extreme bending of light occurs in the vicinity of black holes, where gravity is so strong that light—and everything else that ventures too near—is swallowed. As gas spirals into a black hole, the intense gravitational force field of the black hole accelerates, compressing and heating the gas to millions of degrees—a process best studied with an X-ray telescope.

Such technology was far in the future for Einstein, who did not think that nature would permit the existence of black holes. He turned his attention to the universe as a whole. He made two assumptions: (1) that matter in the universe is distributed more or less uniformly throughout space; and (2) that the radius of the universe is independent of time. In order to keep the universe from collapsing under its own weight, he added another term, which came to be known as "the cosmological constant," to his equations. The cosmological constant acted as a force to keep the universe from collapsing. At this point, as in the case of black holes, Einstein's keen physical intuition failed him. His assumption that the universe has a constant radius turned out to be wrong. It was an understandable mistake. In 1917, when he published his paper on the application of the general theory of relativity to the universe as a whole, virtually everyone believed that the universe was static on a cosmic scale. But Einstein was not virtually everyone—if anyone could think outside the box, it was he. Of course, in the absence of evidence, it was perfectly natural to assume that the universe on a large scale was unchanging. As Einstein is reputed to have said on another occasion, "Everything should be made as simple as possible, but not simpler."

The catch was, a static universe was simpler than was possible. By 1917, Vesto Slipher, an astronomer at the Lowell Observatory in Flagstaff, Arizona, had already gathered evidence that radiation from atoms in a few spiral nebulas was systematically shifted toward longer or redder wavelengths, implying that they were receding from us. It would take another decade of observational work, led in large part by Hubble, before it was established that the nebulas are in fact galaxies much like our Milky Way and that the universe is expanding. Although Einstein apparently did not investigate the possibility at the time, in the years that followed the publication of his general theory of relativity, several scientists did. Willem de Sitter, a Dutch scientist; Alexander Friedmann, a Russian; Georges LeMaître, a Belgian; and Howard Robertson, an American, independently wrote papers showing that the equations of general relativity without the cosmological constant were consistent with a universe that could either collapse or expand, depending on what was assumed about how it all started. Einstein accepted this as the correct description of the universe, and he declared that the introduction of the cosmological constant into the equations of general relativity was "less natural" from a purely theoretical point of view and should be abandoned.

Most astronomers agreed with Einstein, and for roughly the next fifty years, the efforts of cosmologists who worked on such things were directed toward understanding how rapidly space is expanding and fitting their observations into model universes without a cosmological constant. Their primary goal was to learn the age of the universe and to determine whether there is enough matter in the universe to slow down the expansion and turn it around. The idea is that if there is sufficient matter, then the self-gravity of the universe will act against the expansion and cause it to decelerate and perhaps even collapse at some date tens of billions of years in the future. The age of the universe follows from the rate of expansion, and the ultimate fate of the universe from a measure of the rate that the expansion slows.

It was conceptually straightforward but practically very difficult work, as Hubble pointed out. However, thanks to a great telescope, the Hubble Space Telescope, named after the great man, together with work on ground-based telescopes and improved solid-state detectors using charge coupled devices (CCDs), big strides were made in the 1980s and 1990s. Evidence accumulated that the expansion of the universe began somewhere between 10 and 15 billion years ago. So far, so good. Cosmologists acknowledged that the numbers needed tidying up a bit—the age of the universe was still uncertain by 20 percent or so—but the real action moved on to the measurement of the rate of change of the universal expansion. The discovery of the cosmic microwave background radiation in the 1960s convinced most astronomers and astrophysicists that the Big Bang theory was correct: the present universe evolved from a very dense, hot state that existed about 10 to 15 billion years ago. But serious questions remain unanswered.

Chief among these is the uniformity problem. Why is the microwave background radiation so smooth? It is uniform to less than a percent over the entire sky. This uniformity indicates that the matter in the universe must have been distributed extremely evenly less than a million years after the Big Bang. Sound waves or thermal conduction must have smoothed everything out and created a state very near to equilibrium before that point. A simple calculation based on the standard Big Bang model showed that there had simply not been enough time for this to happen. Yet it did happen.

The problem is analogous to one I encountered when teaching a calculus class one summer to a group of bright but underperforming—and even unscrupulous—students. Their answers on a test were remarkably good and remarkably uniform. This suggested two possibilities. Either they were all good students with identical approaches to solving the problems, meaning that I was an extraordinarily good teacher, or there had been surreptitious communication among the good students and the others, meaning that I was not an extraordinary teacher but a poor monitor. Independent evidence suggested the latter. The good students finished the test fairly quickly but did not immediately turn in their papers. Instead, they asked questions for the purpose of distracting their somewhat gullible instructor while the answers were passed around to the rest of the class. By the time the testing period was over, they had more or less uniform answers on the test. That is, the class had come to equilibrium.

The uniformity of microwave background radiation suggests that something similar happened very early in the evolution of the universe. But how? This puzzle was around for about a decade and a half after the discovery of this uniformity, and it was sometimes talked about but was often just filed away as a problem to be dealt with later. Then along came Alan Guth of MIT and other astrophysicists, notably Andrei Linde of Stanford, with their models of the first few billionths of an octillionth of a second in the evolution of the universe. These inflationary universe models postulated that the uniformity problem could be solved if the universe underwent a very brief period of extremely rapid expansion, or inflation, before settling down to the more sedate expansion that astronomers observe in the present epoch.

The universe could have expanded slowly and come to equilibrium—with all regions having approximately the same temperature and density—and then expanded very rapidly for an extremely brief time, after which time the universe was reheated and proceeded to expand in an orderly way for the next 10 billion years or so, to produce the universe as we observe it today.

One clear prediction of the inflationary universe theory was that the curvature of the universe would be zero. The curvature of a sphere decreases in inverse proportion as the radius of the sphere increases. Thus, a golf ball has greater curvature than a basketball, which has greater curvature than Earth, which has greater curvature than the Sun. If, as the inflationary universe model predicted, the universe is enormously larger than that portion of the universe we observe, then the curvature must be close to, and for all practical purposes, zero. In the context of Einstein's theory, the curvature of the universe depends on a quantity called Omega. This number is the ratio of the combined mass-energy density of the universe to a critical value. When the mass-energy density is equal to the critical value, the curvature of the universe is zero and Omega = 1. This solution had great aesthetic appeal, since Omega = 1 seems to be the most natural value for an evolving universe. Any other value seems to require arbitrary fine-tuning.

But there was a problem. Even when astronomers added up all the mass-energy density they could detect by whatever means, including the best estimates of the amount of dark matter in the universe, they came up short. At best, they could account for only 30 or 35 percent of the mass needed to make the curvature of the universe equal to zero.

How to reconcile this dilemma? It could be that the estimates of the mass density of the universe were off by a factor of three. A few scientists suggested, on the basis of the inflationary model and other cosmological problems, such as the age of the universe, that the curvature of the universe is indeed zero. The way to support this argument was to resurrect the cosmological constant, or some unknown energy density. Although the observational evidence for a cosmological constant was at that time sketchy at best, this suggestion was in line with a view suggested years ago by the eminent astronomer Arthur Eddington: "It is also a good rule not to put overmuch confidence in the observational results that are put forward until they are confirmed by theory" (1978).

However, most astronomers took their cue not from Eddington but from Shakespeare, believing that "there are more things in heaven and earth, Horatio / Than are dreamt of in your philosophy." They waited for the data to decide. It did, in the late 1990s, thanks to improvements in technology, which enabled astronomers to use large CCD cameras to survey much larger swaths of the sky for Type Ia supernovas. These surveys demonstrated that the expansion of the universe is accelerating, not decelerating, as almost everyone had expected to find. This astonishing result earned Nobel prizes for Saul Perlmutter of Lawrence Berkeley National Laboratory, Adam Reiss of Johns Hopkins University, and Brian Schmidt of the Australian National University. Their research, which involved two large groups of scientists, one led by Perlmutter, the other with Reiss and Schmidt taking the lead, showed that the amount of acceleration of the universal expansion is consistent with a universe in which Omega = 1, meaning that the cosmological constant, or some other form of energy, must supply roughly 70 percent of the mass-energy density of the universe. Making an analogy with dark matter, the required mass-energy density needed to make Omega = 1 was dubbed "dark energy," by Michael Turner, one of the astrophysicists who had suggested a few years earlier that such a component of cosmic energy—he called it the X-component at the time—could solve the curvature problem. Although astrophysicists still have no idea what dark energy is, they agree its nature is intimately connected with the nature of space itself.

The evidence from the supernovas was convincing, but, to use a phrase popularized by Carl Sagan in *Cosmos*, "Extraordinary claims require extraordinary proof" (1980). In science, such proof usually means independent evidence gathered by independent investigators. If techniques for gathering the evidence differ among the investigators, so much the better. Several different lines of evidence bolster the case that we live in a universe in which dark energy plays a major role. This supporting evidence is especially compelling in that it uses observations from microwave, optical, and X-ray telescopes, and it spans cosmic times from three hundred thousand years after the Big Bang to the present.

The cosmic microwave background radiation—the afterglow of the Big Bang—is observed to be uniform over the sky. One of the greatest achievements in cosmological research has been the detection of tiny temperature variations or fluctuations (at the part-per-million level) by two satellites sensitive to microwave radiation: NASA's Wilkinson Microwave Anisotropy Probe (WMAP), which operated from 2001 until 2010, and the European Space Agency's Planck telescope (2009–13). These temperature fluctuations (see figure 8, page 19) are caused by clumps of matter that have a slightly higher or lower density than the average. The growth of these clumps depends on such things as the expansion rate of the universe and the speed at which sound waves travel, which in turn depend on the mass-energy density and composition of the universe. Oscillations due to these sound waves show up as a subtle pattern in a map of the cosmic microwave background. The data from WMAP and Planck both show that the amount of dark energy required is consistent with the results of supernova studies. Furthermore, they require an age for the universe of 13.7 billion years, which eliminated another potential headache: the ages of the oldest stars had been confidently determined to be greater than 11 billion years, in conflict with some models of the universe.

The clumps detected by WMAP and Planck grew over the next 10 billion years to form the galaxies of today, so the pattern of the clumps remains imprinted on the distribution of matter and shows up in the distribution of galaxies formed hundreds of millions of years later. One of the primary goals of the Sloan Digital Sky Survey is to determine the present-day distribution of galaxies in the universe. The Sloan Survey team looked at a sample of 46,748 luminous red galaxies over 3,816 square degrees of sky (approximately 5 billion light-years in diameter) and out to distances of about 5 billion light-years (see figure 5, page 11). The lumpiness in the distribution of galaxies was consistent with the expectations from WMAP and Planck and provides another line of evidence.

Chandra's observations of galaxy clusters have provided independent confirmation. Steve Allen of Stanford University and his colleagues used Chandra to study clusters of galaxies at distances corresponding to 1 to 8 billion light-years. These data span the time when the universal expansion slowed before speeding up again.

Because galaxy clusters are the largest collapsed objects in the universe, the relative amount of baryonic matter and dark matter in clusters should be a good indication of the relative amounts of such matter in the universe as a whole. Observations and computer simulations indicate that the relative amount of mass in hot gas and dark matter should be approximately constant over time. What makes this interesting is that determination of this ratio, called the gas fraction, for extremely distant clusters depends on assumptions about the curvature of space, which in turn depends on the amount of dark energy. Measurements of the gas fraction for a cluster depend on what is assumed about the amount of dark energy. The numbers on the amount of dark energy can then be adjusted until they give the right answer for the gas fraction. The result then provides a measurement of the amount of dark energy. It is similar, in a way, to balancing a scale. You put a cabbage on one side and a known weight on the other side, keep adding weights until you get a balance, and you can then compute the weight of the cabbage.

Allen's group analyzed Chandra observations of a sample of forty-two galaxy clusters distributed over a wide range of distances. They found that the amount of dark energy required to explain their observations is consistent with the measurements from supernovas and the cosmic microwave background.

Observations of extremely distant supernovas indicate that the expansion of the universe was slowing down due to the gravitational pull of all the matter in the universe for about 9 billion years. Dark energy did not become dominant until about 5 billion years ago, when the expansion of the universe switched from deceleration to acceleration.

The early dominance of gravity was critical in that it allowed galaxies and clusters of galaxies to "condense out" of the expanding gas. The push-pull struggle between the push of accelerating expansion and the decelerating pull of gravity determines the expansion rate of the universe. Since galaxy clusters are still in the process of forming, the amount of dark energy also affects the speed at which they form and grow. In the absence of cosmic acceleration, the universe would be expanding at only half the present rate. More galaxies would have merged, and more galaxies would have been pulled together into groups, and more groups would have merged to form clusters.

A study of how fast galaxy clusters actually did form should thus reveal the expansion history of the universe, and thereby its dark energy content. The best way to make such a study is with Chandra observations of galaxy clusters. These observations can be used to determine the masses of clusters out to very large distances, and thereby to search for evidence of cosmic acceleration produced by dark energy.

Alexey Vikhlinin of the Harvard-Smithsonian Center for Astrophysics and his colleagues did just that. In a research paper published in 2009, they used Chandra data to determine the mass of eighty-six galaxy clusters (including Abell 2052; figure 14, page 42) at distances ranging out to 7 billion light-years and showed that the observed number of massive galaxy clusters was about five times less than was predicted in the absence of cosmic acceleration. The rate of formation of massive galaxy clusters has slowed down over the past few billion years in a manner consistent with expectations in an accelerating universe.

"This result could be described as 'arrested development of the universe,'" said Vikhlinin. "Whatever is forcing the expansion of the universe to speed up is also forcing its development to slow down." Multiple lines of evidence now show that dark energy is the missing piece of the puzzle that makes it possible to bring many observations about the universe into concordance. In fact, cosmologists often refer to the collection of facts that describe our present knowledge as the concordance model (figure 15). According to a detailed analysis published by the Planck Collaboration in 2015, nearly 5 percent of the mass-energy in the universe is normal matter consisting of protons, neutrons, and electrons; 26 percent is dark matter; and 69 percent is dark energy. The concordance model can account for the existence and structure of the cosmic microwave background; the large-scale structure in the distribution of galaxies; the abundances of the light elements: hydrogen (including deuterium), helium, and lithium; and the accelerating expansion of the universe.

The concordance model represents the impressive progress that has been made over the past two decades in understanding the overall shape and makeup of the universe. But if the devil is in the details, then the concordance model definitely has some bedeviling details to deal with. One is: What exactly is that 95 percent of the universe that we call dark? It is still of completely unknown form, yet it is crucial to the existence of the universe and will determine its destiny.

That's really not a detail when you think about it. Instead, it's a major problem.

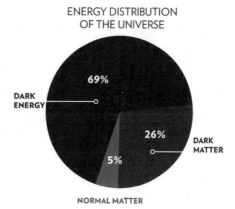

ENERGY DISTRIBUTION OF THE UNIVERSE

69%

DARK ENERGY

26%

DARK MATTER

5%

NORMAL MATTER

15 | ENERGY DISTRIBUTION OF THE UNIVERSE

The total energy distribution of the universe as per the concordance model. Dark energy is estimated to contribute 69 percent of the universe's energy, dark matter 26 percent, and normal matter 5 percent. Only normal matter can be directly detected with telescopes.

16 | ABELL 383

A composite image of the Abell 383 galaxy cluster, about 2.3 billion light-years from Earth. Chandra X-ray data are shown in purple, and optical data from the Hubble Space Telescope, the European Southern Observatory's Very Large Telescope, and the Sloan Digital Sky Survey are in blue and white. Abell 383 is one of the forty-two galaxy clusters that researchers have used to show that galaxy clusters' observed growth is consistent with Einstein's theory of gravity. The image is about 4.2 million light-years on a side.

6

A universe dominated

by a cosmological constant is

a strange place to live.

David Spergel, "The Dark Side of Cosmology," 2015

WHAT IS DARK ENERGY?

There are two basic models of dark energy. One holds that it is energy associated with empty space (vacuum energy) and is constant throughout space and time, equivalent to the so-called cosmological constant; the other argues that it is an energy field that varies over space and time—and it is also known as a scalar field or quintessence. The recently discovered Higgs field, which determines the mass of subatomic particles, is such a field. So is the inflaton field that supposedly drove the exponential or inflationary expansion of the universe for the first fraction of a second of its existence. Coming up with a theory about dark energy has proved elusive.

Vacuum energy is the most straightforward explanation for dark energy. So far, the various probes of dark energy have been consistent with a constant value for vacuum energy. However, the physical basis for vacuum energy is a mystery. Einstein introduced the cosmological constant into his equations for general relativity in order to produce a static universe, but he dropped the constant as unnecessary when Edwin Hubble and others showed that the universe is expanding.

With the development of quantum mechanics, scientists realized that the Heisenberg uncertainty principle allowed particles to blink into and out of existence on extremely short time scales (about a trillionth of a nanosecond for electrons), so that empty space, or the vacuum, is not truly empty. These so-called virtual particles produce a shift of energy levels in hydrogen atoms and in particle masses, called the Lamb shift, after Willis Lamb, one of its discoverers, along with Robert Retherford, in 1947. In 1955, Lamb was awarded the Nobel Prize in physics for this discovery.

Attempts to estimate the energy density associated with the quantum vacuum led to the absurd result that vacuum energy density should be at 60 to 120 orders of magnitude larger than is observed. No satisfactory explanation for resolving this enormous discrepancy has been put forward. Advances in understanding the nature of elementary particles, perhaps stimulated by discoveries with the Large Hadron Collider at the European Organization for Nuclear Research (CERN), may shed light on vacuum energy in the near future.

The takeaway is that dark energy density, though dominant on a cosmic scale of billions of light-years, is small on any scale of millions of light-years or less. This is good for us, because it keeps the room from flying apart, but it requires exquisite fine-tuning to explain why it is so small.

Another idea is that the universe is but one of many universes, each with different values of vacuum energy. Only those universes with a small value of vacuum energy would allow formation of stars and galaxies. In essence, we could not exist in a universe that has a large value of vacuum energy, so the fact that we exist and are worrying about it means that it must be small, so we shouldn't worry about it. It just is. Still, scientists do worry, and well they should. As Bruno Rossi, one of my early mentors, once said to me concerning some perplexing observations that seemed relatively unimportant: "As long as we don't understand them, we don't know how important they are."

Vacuum energy, or the cosmological constant, is constant in space and time, as its name implies. A more general possibility is that the accelerated expansion is driven by a force field that is called a scalar field, or quintessence. One class of such models assumes that the scalar field energy density tracks the energy density of radiation and matter at very early times and then comes to dominate the energy density of the universe at later times. This tracking property could provide an explanation for a peculiar cosmic coincidence: Why are the energy densities of dark matter and dark energy about the same? Recall that scientists were excited when they discovered dark energy because it made Omega, a measure of the total energy density of the universe, equal to unity. And they should have been, because it solved a host of problems. But why does dark energy contribute roughly 70 percent, and dark matter roughly 25 percent, with about 5 percent left over from baryonic matter? What is so special about these ratios? Why hasn't one form of mass-energy density gained the upper hand at the expense of the others? Is it because we live in some special epoch in the evolution of the universe? Tracking models provide a prescription but no real explanation for why these energy densities are as they are.

Many versions of scalar fields have been proposed, but as yet none has emerged as a favorite. In particular, as with vacuum energy, no satisfactory physical explanations exist for why the strength of the scalar field is so small, or for the relation, if any, between dark matter and dark energy.

An important goal of future research is to distinguish between vacuum energy and scalar fields as dark energy candidates. The most promising way to do this is to determine the exact relation between the density and the pressure of dark energy. Dark energy has the peculiar property that its pressure is negative. Normally, when a gas expands, its energy decreases. Dark energy does just the opposite: the amount of dark energy keeps increasing as the universe expands. Conservation of energy goes out the window on a cosmic scale, but that happened a long time ago, with the discovery of the expanding universe. Details of the formation of clusters of galaxies and other studies depend on the precise relationship between the pressure and the energy density, so observations can set limits on this relation and determine whether it is varying with time. To date, all observations have been consistent with a constant ratio of pressure and energy density, but other values, as well as variation with time, are not ruled out.

All the evidence for dark energy uses Einstein's general theory of relativity to interpret its observations. Might it be that cosmic acceleration is a sign, not of dark energy, but that the theory of gravity needs to be modified for extremely large distance scales?

Models that seek to explain cosmic acceleration by changing the law of gravity must pass several observational tests. They must not change the conditions in the early universe so much as to spoil the successful calculation of the amount of helium produced by hydrogen nuclear fusion reactions in the first few minutes, or the sizes of fluctuations in the cosmic microwave background, which reveals the state of the universe a few hundred thousand years after the Big Bang. The models must also account for the observed rate of growth of clusters of galaxies billions of years after the Big Bang. This latter test involves Chandra.

Observations of the growth of galaxy clusters can be compared using the predictions of theories that require modifications of the general theory of relativity. As discussed earlier, such a test was used to establish the existence of dark energy, assuming Einstein's general relativity theory is correct. But what if it is not correct? After all, desperate times call for desperate measures. One such theory is called $f(R)$ gravity, where R is the curvature of the universe and $f(R)$ is an algebraic function that describes the deviation of the theory from the general theory of relativity. In $f(R)$ theories, the acceleration of the expansion of the universe is not driven by an exotic form of energy but by a modification of the gravitational force. Such modifications are predicted in some versions of string theory and in other theories that attempt to unify quantum physics and gravity. This approach is similar to Mordecai Milgrom's attempt to do away with dark matter, and as with Milgrom's, it runs into trouble when dealing with physics on the scale of galaxy clusters.

The catch, with the theory of extra force, is that it implies an extra energy density and slows down the expansion ever so slightly. It is possible to adjust the magnitude of the hypothetical extra force to avoid the problems in the first few hundred thousand years after the Big Bang, but over the billions of years needed for galaxy clusters to form, it takes its toll. This is why X-ray telescopes such as Chandra are proving to be such sensitive tools for observing galaxy clusters and studying the evolution of the universe.

Fabian Schmidt of the Max Planck Institute in Germany, along with Vikhlinin of CfA and Wayne Hu of the University of Chicago, used Chandra observations of forty-nine galaxy clusters to determine the growth of galaxy clusters over time. This data, together with computer simulations and data on fluctuations in the cosmic microwave background and the large-scale distribution of galaxies and supernovas, show that $f(R)$ modifications to Einstein's theory are not detected to a level of less than a hundredth of a percent. This limit, which applies to distances ranging from 130 million light-years to several billion light-years, is a hundredfold improvement on limits set with pre-Chandra data.

In a second study, David Rapetti of the University of Copenhagen and his colleagues used X-ray observations from Chandra and the German Roentgensatellite (ROSAT), which operated from 1990 to 1999. The X-ray data showed how rapidly galaxy clusters have grown over cosmic time. These results, together with optical data and data on the cosmic microwave background, were used to test the predictions on the growth of clusters based on Einstein's theory (figure 16, page 54). Data from geometrical studies, such as distances to supernovas and galaxy clusters, were also incorporated. Nearly complete agreement between observation and theory was found. This agreement argues against any alternative gravity models with a different rate of growth. In particular, a model with five dimensions predicts a slower rate of cluster growth than Einstein's theory does because gravity is weakened on large scales as it leaks into an extra dimension. This model is in conflict with Chandra's observations of galaxy clusters' rate of growth.

As Roger Blandford (2015) of Stanford University put it in a review written on the occasion of the centennial of Einstein's general theory of relativity, "There seems to no need for baroque variations and empirical corrections over its domain of applicability." In other words, Einstein's theory is still the best game in town.

An alternative to modifying gravity is to assume that the Milky Way, in which we live, is located near the center of a very large, nearly spherical region that has a lower density than the average density of the universe. It remains to be seen whether such a model can explain all the evidence for cosmic acceleration.

In the next decade, telescopes will map the large-scale structure of the universe over the past 10 billion years and trace the distribution of matter over much of the sky. These observations should provide more stringent tests of general relativity and further constrain the properties of dark energy.

In the meantime, we are left to contemplate a dark future. Dark energy has the nightmarish property that, as the universe expands, it creates more dark energy. How much more is uncertain, but such repulsive behavior could eventually get badly out of hand. The data still allow for the possibility that, several billion years from now, the ever-increasing dark energy could pull space apart so rapidly that galaxy clusters would become unbound, followed by the disintegration of individual galaxies such as the Milky Way, then solar systems, until finally matter itself would be shredded by accelerating space. The universe would end in a "big rip." Downright scary, befitting a horror story by Edgar Allan Poe.

17 | GALAXY CLUSTER COLLISION

A composite X-ray (Chandra) and optical (Hubble) image showing the huge galaxy cluster MACSJ0717.5+3745 (MACSJ0717 for short), about 5.4 billion light-years from Earth. Four galaxy clusters have collided here, the first time such a phenomenon has been observed in detail. The Hubble image shows the individual galaxies (labeled A through D), and the Chandra image shows the hot gas, which has a mass roughly five times that of the galaxy clusters. The Chandra image is color-coded to show the range of temperatures in the gas, from 60 million degrees Celsius (reddish purple) to more than 200 million degrees Celsius (blue). The arrows indicate three of the clusters' approximate movement in a direction perpendicular to the line of sight. The direction of the clusters' motion is about parallel to the direction of the filament (left), a 13-million-light-year-long stream of galaxies, gas, and dark matter.

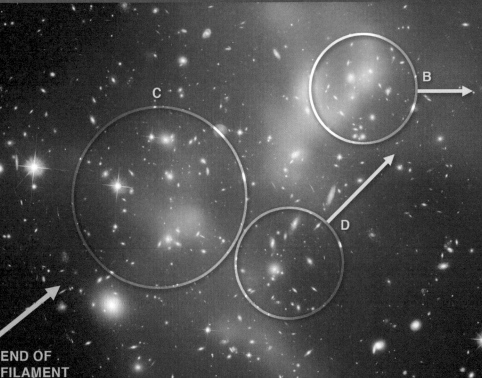

A

B

C

D

END OF
FILAMENT

7

So we paint the following picture of
how clusters and their filamentary bridges
weave a web.

J. Richard Bond, Lev Kofman, and Dmitri Pogosyan, "How Filaments of
Galaxies Are Woven into the Cosmic Web," 1996

THE COSMIC
WEB

At scales of millions of light-years, the distribution of matter in the universe is not uniform. Instead, it forms an intricate pattern known as the cosmic web, a phrase coined by Richard Bond of the University of Toronto and colleagues Lev Kofman of the University of Hawaii and Dmitri Pogosyan of the University of Toronto. Their pioneering 1996 paper described how such a structure could arise naturally as a result of the gravitational collapse of primordial clumps that existed in the early universe.

As discussed in chapter 2, the presence of the cosmic web can easily be seen in the distribution of galaxies (see also figure 3, pages 6–7). The cosmic web consists of the largest structures in the universe, with massive galaxy clusters as nodes that are interconnected through an intricate web of filaments and sheets of tenuous gas and galaxies, with nearly empty regions called voids taking up most of the volume. Under the pull of gravity, matter flows from voids into sheets, from sheets into filaments, and from filaments into clusters of galaxies. Each of these components shows substructure in the form of galaxies that took shape before the larger features of the cosmic web did.

Computer simulations of the formation of the cosmic web indicate that most of the mass in the universe is in the filaments and that it will flow along these rivers of gravity and eventually wind up in clusters of galaxies. Bond and his colleagues showed that just as the uneven pull of the Moon on the near and far sides of Earth pulls it slightly out of round and produces tides, primordial gravitational tidal forces create the cosmic web. First matter collapses to form walls, or sheets of gas and galaxies, then the sheets collapse into filaments, and finally the matter flows into clusters. In the two decades since the groundbreaking work of Bond and his colleagues, the sophistication of simulations of the evolution of the cosmic web has increased with the growth of computing power. These simulations now include finer spatial resolution and physical processes such as the energetic feedback from the formation of stars, supernova explosions, and energy generated by matter falling into black holes. The most ambitious effort to date is the Illustris project, a multinational effort led by Mark Vogelsberger of MIT. The Illustris simulations are run on supercomputers in France, Germany, and the United States. The largest so far was run on 8,192 computer cores and took 19 million central processing unit (CPU) hours, the equivalent of one computer CPU running for 19 million hours, or about two thousand years. Figure 3 shows the evolution of the dark matter density from an age of 1.5 billion years (redshift $z = 4$) up the present ($z = 0$). The bright patches at the intersections of filaments represent galaxy clusters, each of which contains several hundred, and in some cases, several thousand galaxies, while each galaxy contains billions of stars.

Figure 17, page 60, shows the final stages of this process, as captured by Chandra. A massive galaxy cluster, MACSJ0717.5+3745 (MACSJ0717 for short), located about 5.4 billion light-years from Earth, is the product of the collision of four separate galaxy clusters. This is the first example detected of such an enormous, complex collision. Repeated collisions are caused by galaxies, hot gas, and dark matter pouring into the cluster in a filament that stretches over 13 million light-years. Collisions between gas clouds cause the hot gas to slow down. However, the galaxies, which are mostly empty space, do not slow down as much and so move ahead of the gas. The speed and direction of each cluster's motion can be estimated by studying the offset between the average positions of the galaxies and the peaks in the hot gas. The gas, which reaches temperatures exceeding 200 million degrees Celsius, is color-coded, with the hottest gas in blue and the coolest gas in reddish purple. An optical image from the Hubble Space Telescope shows the galaxies.

Figure 17 shows the galaxies in the four different clusters—labeled A, B, C, and D—involved in the collision, plus the direction of motion for the three fastest-moving clusters. The length of each arrow indicates the approximate speed in a direction perpendicular to the line of sight. Note that the direction of motion of the clusters is roughly parallel to the direction of the filament, which is a huge stream of galaxies, gas, and dark matter. Data from Keck Observatory was used to calculate the speed of the clusters along the line of sight and to figure out the three-dimensional geometry and dynamics of the collision.

The cooler (redder) region of gas at the lower left of cluster D has likely survived since before the collision. Cluster A is likely falling back into the main cluster after passing through it once in the opposite direction. Both clusters probably originated from the filament. Cluster B, however, has a much faster speed than the others along the line of sight, and its origin is unclear. It may have fallen along the outer edge of the filament, causing its infall trajectory to curve, or it might be falling in along another, smaller filament. The good alignment between the galaxies and hot gas for cluster C, along with its motion compared to MACSJ0717 as a whole, makes cluster C a good candidate for the core of the main cluster. The accompanying wide-field X-ray image shows the approximate position of the large-scale filament leading into the cluster. Material flows along the filament into the cluster.

MACSJ0717 provides perhaps the best example of matter, both dark and baryonic, flowing from a filament into a cluster. However, detection of the baryonic matter in the filaments of gas among the clusters has been challenging. Astronomers know it must be there. The synthesis of helium from hydrogen in the first few minutes of the Big Bang gives a firm figure for the number of protons per cubic meter. Running those numbers forward 13.7 billion years to the present, and taking into account the expansion of the universe, yields about a hundred protons every five hundred cubic meters, roughly the volume of a two-thousand-square-foot house with eight-foot ceilings. This calculation is an average over the universe, most of which is nearly empty space.

Here is the problem: An estimate of all the matter in all the stars and gas clouds in all the galaxies, and in all the hot gas in all the galaxy clusters, still yields only 50 percent of the amount implied from the Big Bang calculation. Where is the rest of it? The best bet is that it is in the cosmic web, like a fog lying along valleys carved by the rivers of gravity. Several teams of scientists have undertaken a search for it, using Hubble and Chandra. Computer simulations that show the development of the cosmic web indicate that the gas in the filaments should have

18 | X-RAYING THE COSMIC WEB

A schematic of X-rays of various energies emitted by a distant quasar, shown as waves of colored light. As they pass through a cloud of intergalactic gas, which is the white, filamentary structure, X-rays in a narrow range of energies (in this sketch, the ones in yellow) are absorbed preferentially by oxygen atoms in the intergalactic gas cloud.

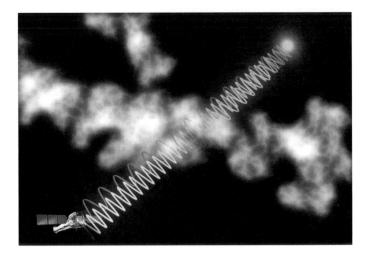

a temperature in the range of 100,000 to 10 million degrees Celsius. This gas, which goes by the acronym WHIM, for warm-hot intergalactic medium, should radiate primarily at X-ray energies. This radiation is produced when an electron strikes an atom, and it is expected to be very feeble because the atoms in the filaments are rare, and collisions that produce X-rays are extremely rare. In some instances, it is much more likely that X-rays from a bright source behind a filament will be absorbed by an atom in the filament, thus revealing the presence of the WHIM.

The Hubble Space Telescope has found good evidence for the WHIM in the form of telltale absorption detected in light from distant quasars. These extremely luminous sources of radiation are produced by matter pouring into supermassive black holes. The quasars Hubble observed in this study are 10 billion or more light-years away, so we see them as they were when the universe was less than 4 billion years old. As light from a quasar travels toward Earth, it may traverse one or more intergalactic clouds. Some of the light will be absorbed and result in a dark line, or shadow, in the spectrum of the quasar. Hubble has detected these shadows at ultraviolet energies. However, an analysis shows that these observations detect only about 10 percent of the expected WHIM. The likely explanation for this is that the WHIM has a high temperature of about a million degrees Celsius or more. If so, the outer electrons of the vast majority of the oxygen ions in the WHIM will be stripped away and can be detected only through their ability to absorb X-rays.

Observations of quasars by groups from MIT, the Harvard-Smithsonian Center for Astrophysics (CfA), UC Irvine, and Ohio State University have revealed various parts of the hot gas system. The best evidence to date for the hot component of the WHIM (figures 18–19) is from a detection by Taotao Fang of UC Irvine and colleagues of evidence for absorption of hot gas in the spectrum of an X-ray source 4 billion light-years from Earth. The gas producing the absorption is in the Sculptor Wall (figure 19), a sheetlike arrangement of galaxies about 400 million light-years from Earth that stretches across tens of millions of light-years. The amount of the absorption, which was detected by both Chandra and XMM-Newton, is consistent with expectations based on the cosmic web computer simulation.

So the WHIM is indeed out there, as the computer simulations predicted, moving along an intergalactic superhighway and transporting matter into the giant galaxy clusters, the "urban centers" where the action is.

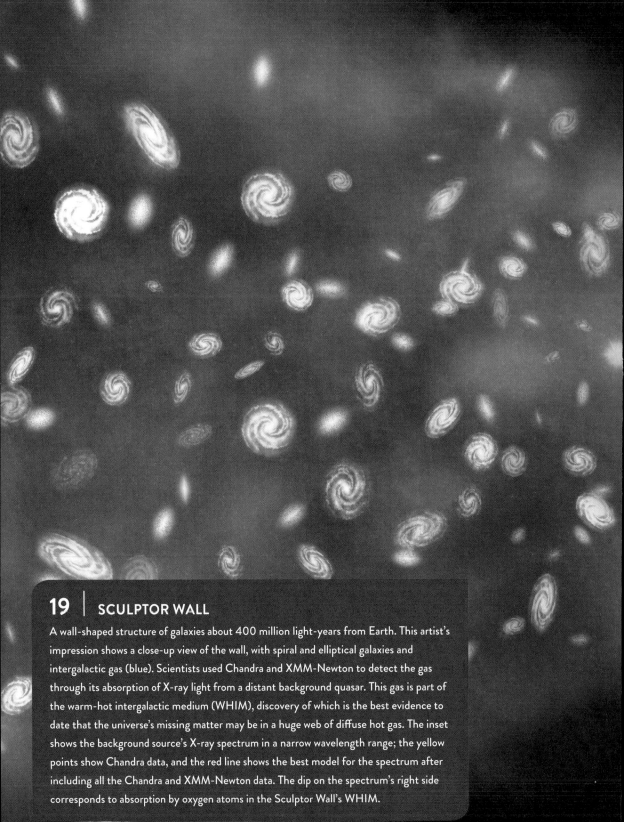

19 | SCULPTOR WALL

A wall-shaped structure of galaxies about 400 million light-years from Earth. This artist's impression shows a close-up view of the wall, with spiral and elliptical galaxies and intergalactic gas (blue). Scientists used Chandra and XMM-Newton to detect the gas through its absorption of X-ray light from a distant background quasar. This gas is part of the warm-hot intergalactic medium (WHIM), discovery of which is the best evidence to date that the universe's missing matter may be in a huge web of diffuse hot gas. The inset shows the background source's X-ray spectrum in a narrow wavelength range; the yellow points show Chandra data, and the red line shows the best model for the spectrum after including all the Chandra and XMM-Newton data. The dip on the spectrum's right side corresponds to absorption by oxygen atoms in the Sculptor Wall's WHIM.

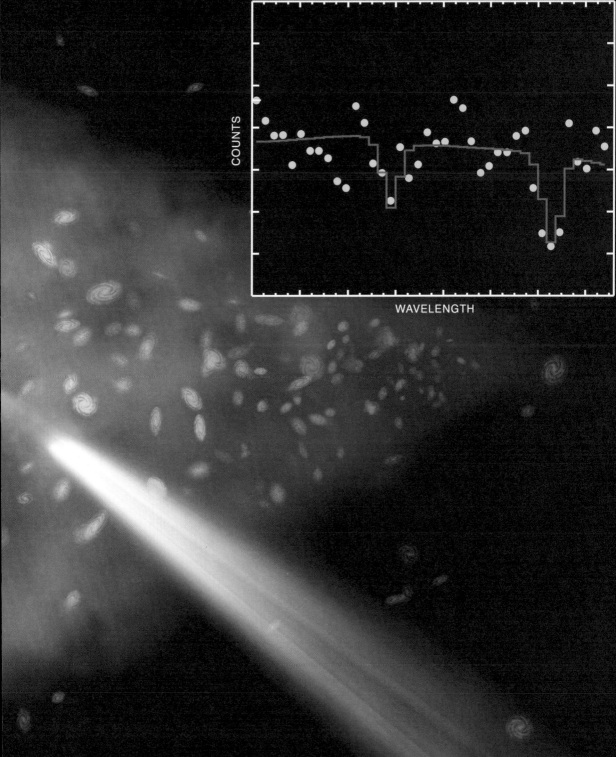

COUNTS

WAVELENGTH

PART II

THE

BAD

20 | **MATTER FALLING ONTO A BLACK HOLE**

An illustration showing a black hole pulling gas from a massive blue companion star. The gas forms a rotating disk (in red and orange) around the black hole. Some gas falls into the black hole, and some is blown away from the black hole in the form of an energetic jet.

21 | GAS IN AN ACCRETION DISK AROUND A BLACK HOLE

A schematic: As the gas circles and descends toward the black hole, frictional forces heat the gas to millions of degrees Celsius. Just before it disappears beyond the event horizon into the black hole, the gas gets so hot that it puffs up to form a crown, or corona, around the black hole. In some cases, magnetic forces combine with gravity to drive jets of particles away from the black hole at extremely high speeds.

8

We shall find

A pleasure in the dimness of the stars.

Samuel Taylor Coleridge, "The Nightingale," 1798

TAKING PLEASURE
IN THE DIMNESS
OF STARS

In astronomy, a key challenge is to explain the unexpected brightness of an object—for example, the central region, or nucleus, of a galaxy that emits intense radiation, or an anomalously luminous supernova. Sometimes, though, the unexpected dimness of objects can lead to important discoveries. One is the finding that distant supernovas are dimmer than predicted, a discovery that suggests that a mysterious "dark energy" dominates our universe.

Ever since Newton came up with his theory of gravity more than three hundred years ago, astronomers have speculated that some stars might have collapsed and become dim—indeed, dark—because their gravity was so strong that nothing, not even light, could escape. In the past forty years, astrophysicists have amassed an impressive body of evidence for the existence of dark stars, or black holes, as they are now called. They look for an object in which the concentration of mass is so great that the mass cannot be stabilized against gravity by nuclear

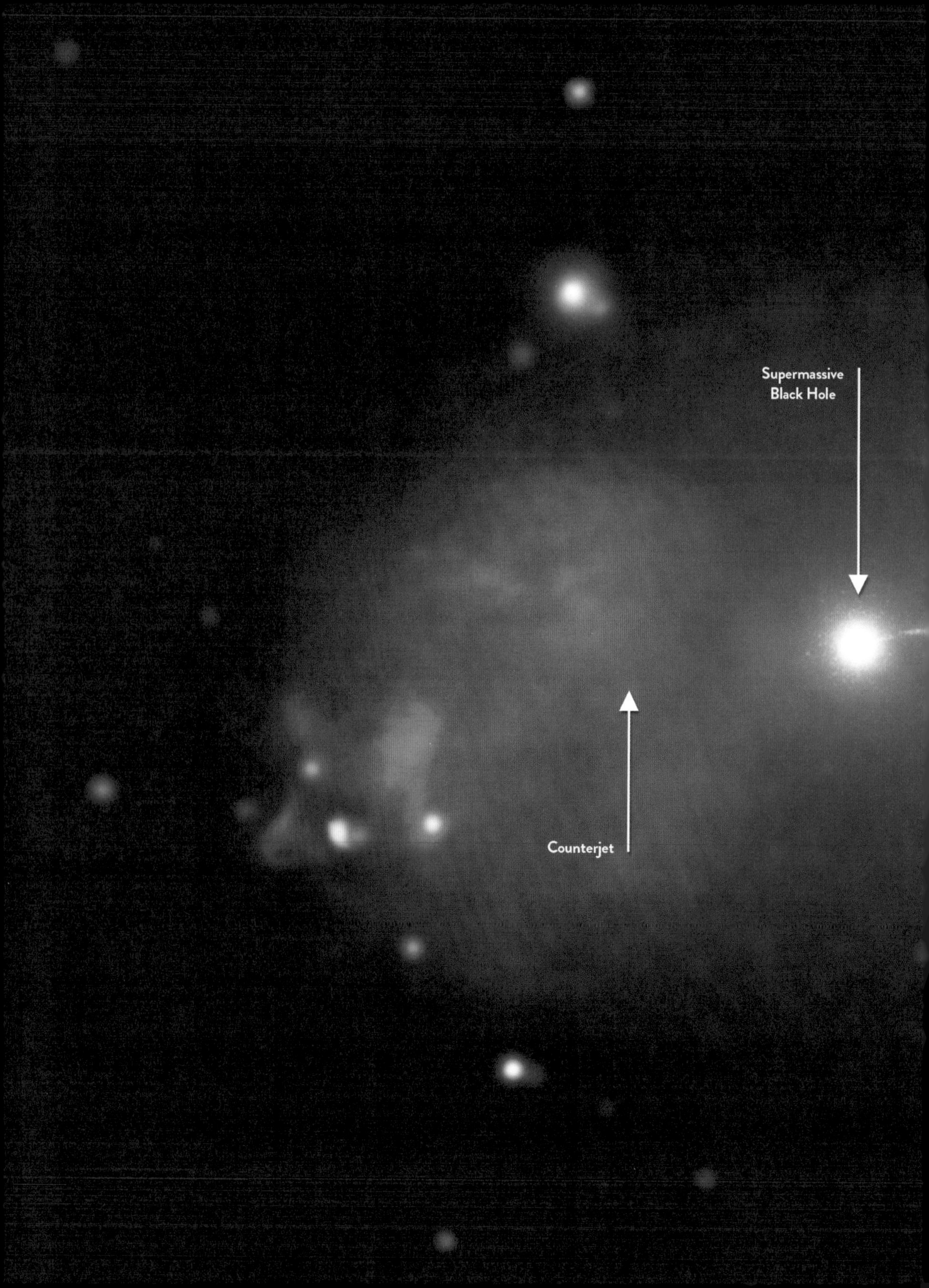

Supermassive Black Hole

Counterjet

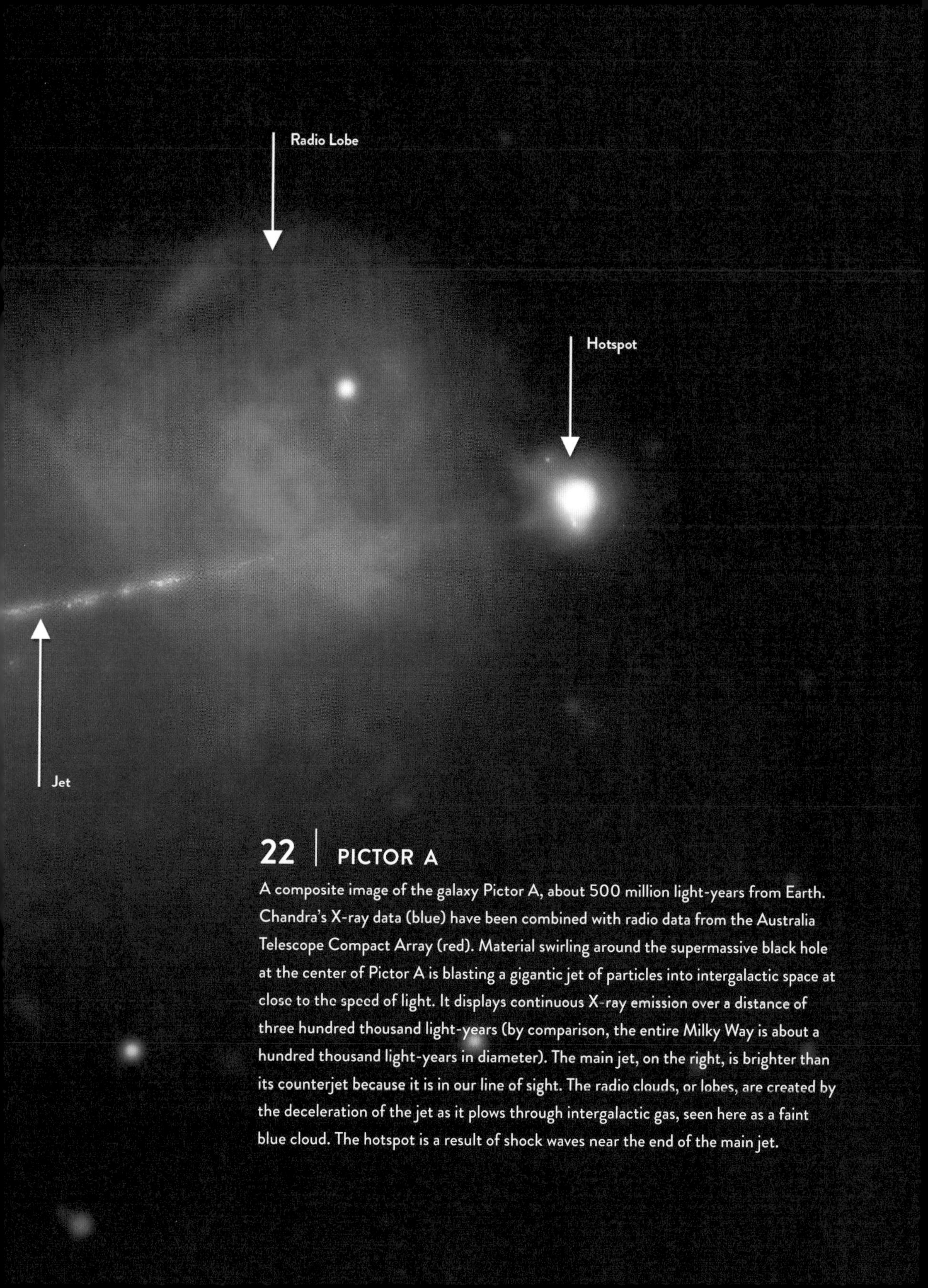

Radio Lobe

Hotspot

Jet

22 | PICTOR A

A composite image of the galaxy Pictor A, about 500 million light-years from Earth.
Chandra's X-ray data (blue) have been combined with radio data from the Australia
Telescope Compact Array (red). Material swirling around the supermassive black hole
at the center of Pictor A is blasting a gigantic jet of particles into intergalactic space at
close to the speed of light. It displays continuous X-ray emission over a distance of
three hundred thousand light-years (by comparison, the entire Milky Way is about a
hundred thousand light-years in diameter). The main jet, on the right, is brighter than
its counterjet because it is in our line of sight. The radio clouds, or lobes, are created by
the deceleration of the jet as it plows through intergalactic gas, seen here as a faint
blue cloud. The hotspot is a result of shock waves near the end of the main jet.

forces or any other known forces. Einstein's general theory of relativity then requires that the object must collapse to form a black hole.

A collapsed star with a mass less than about three times the mass of the Sun can be stabilized as a neutron star by nuclear forces. If the mass is greater than this critical mass, the object is presumed to be a black hole. Although the theory for how massive stars end their lives is still incomplete, astrophysicists are fairly confident that several million, maybe even several hundred million, stars must have formed black holes in our Milky Way galaxy alone.

But black holes do not reveal their presence easily. In principle, you can search for them through their gravitational effects on a companion star if they are in a double (or binary) star system. But that is not simple in a galaxy with many billions of double star systems. Fortunately, the richness of nature helps. Some fraction—probably about half—of all black holes are in a binary star system, and of these, a much smaller fraction are in a binary system in which the stars are so close together that material is pulled from the companion star into the black hole. As this matter flows toward the black hole, it settles into an area called an accretion disk, where the combination of extreme gravity and frictional forces heats the gas to millions of degrees (figure 21, page 70). The hot gas radiates X-rays at a sufficiently strong level to be observed with an X-ray telescope.

All that is really known at this point is that a dark star is present in the system. It could be a neutron star (figure 23). In fact, most of the so-called X-ray binary sources contain neutron stars, not black holes. Further measurements with optical telescopes are needed to confirm that the mass of a dark star is greater than three times the mass of the Sun. If so, then it must be a black hole, if Einstein's general theory of relativity is correct. This is how the first black hole candidate, Cygnus X-1, was discovered in 1972. Since then, about two dozen stellar black hole candidates have been identified in this way. I say "candidates" because all we really know is that the collapsed objects are too massive to be neutron stars. Maybe they are objects made of some of heretofore undiscovered type of matter that can withstand the otherwise overwhelming pull of gravity. As such, they would be evidence of new physics, so this possibility, however remote, cannot be ignored.

For definite proof of "black holeness," scientists must demonstrate that a candidate black hole possesses an event horizon, a boundary in space-time that acts like a one-way membrane separating the black hole from the rest of the universe. Matter, radiation, and energy can fall into the black hole from beyond the event horizon, but nothing, not even light, can get out.

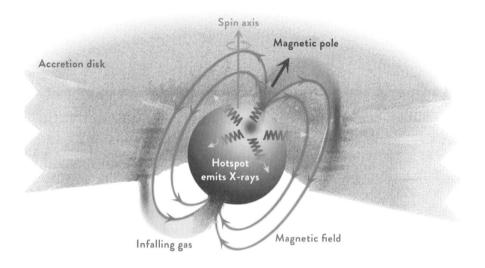

Spin axis

Magnetic pole

Accretion disk

Hotspot
emits X-rays

Infalling gas

Magnetic field

23 | MATTER FALLING ONTO A NEUTRON STAR

Matter falls onto a neutron star from an accretion disk around the star. The strong magnetic field of the neutron star channels the infalling matter to the magnetic poles, where the field is weakest. The accretion onto the star produces a hotspot, and as the star rotates, this hotspot comes into and out of view, producing pulsations in the observed X-ray emission.

When viewed from typical distances of thousands of light-years, the shadow of the approximately 9.3-mile radius—also called the gravitational radius—of the event horizon of a black hole with a mass ten times that of the Sun is much too small to be imaged by telescopes. However, thanks to the ingenuity of astrophysicists, indirect detection of an event horizon might be possible, even feasible.

The important feature of the event horizon is that it is not a surface in the usual sense of the word, such as the surface of the ocean. A particle encountering the event horizon would not collide with anything; it would just coast right through as though nothing were there, because nothing is.

One method to prove that a collapsed massive object has an event horizon and is by implication a black hole is to show that it does not have a surface. This is possible because X-ray binaries have provided a number of excellent cosmic laboratories to test whether the compact objects in a binary system have surfaces and not an event horizon.

Ramesh Narayan and Jeffrey McClintock of the CfΛ have outlined several ways to use X-ray binaries to search for event horizons. Three of these methods contrast the differences in the X-radiation emitted by gas falling onto the surface of a neutron star as compared with gas diving through a black hole event horizon.

One way to distinguish between neutron stars and black holes is the presence or absence of so-called Type I X-ray bursts. When gas pulled from a companion star settles onto the surface of a neutron star, it forms a dense, hot layer. If this layer reaches a critical mass, it becomes a thermonuclear bomb that blows away the accreted material and emits a burst of X-rays. The same process should occur for gas falling onto any compact object that has a surface instead of an event horizon. Yet, although thousands of bursts have been observed from neutron stars, not one has been observed from a black hole candidate. The lack of X-ray bursts from black hole candidates is a strong argument against the existence of gravastars—a portmanteau for a hypothetical "gravitational vacuum condensate star," that is, a dense ball of matter kept inflated by a core of dark energy—and other exotic alternatives that involve a modification of Einstein's theory, because these alternative objects would have surfaces.

Another potentially powerful test for the existence of black hole event horizons comes from the study of X-ray novas. These systems, which are a subclass of X-ray binaries, tend to flare up and become extremely bright, typically for a few months, and then fade away for years and even decades before flaring up again. The explanation for this behavior is that the hot disk of gas around the compact object disappears during the quiescent period. In this state, some gas continues to flow toward the compact object at a very low rate, producing a weak source of X-rays. Because the inflowing gas produces additional X-radiation when it crashes into the surface of a neutron star, quiescent neutron stars should be much brighter than a black hole candidate, which has no surface.

Observations show that X-ray binary systems with similar orbital periods are likely to have similar flows of infalling gas. Using this information, Narayan, together with Jeffrey McClintock and Michael Garcia of CfA, proposed that a comparison of the quiescent X-ray luminosities of neutron star novas and black hole novas could provide a test for the existence of black hole event horizons. Neutron stars and black hole candidates in quiescence should have comparable X-ray luminosities, if both types of objects have surfaces.

The data show that quiescent neutron star novas are one hundred or more times brighter than their black hole counterparts, as is to be expected if black holes have event horizons. This result has been considered strong evidence for the existence of black hole event horizons.

The case for stellar-mass black holes is strong and is supported by several independent lines of evidence. However, in a report discussing stringent new limits set by Chandra on X-radiation from a source named 1H1905+000 (all the numbers refer to the source's location), Peter Jonker of CfA and Utrecht University in the Netherlands and his colleagues argue that these limits may be a show-stopper. They conclude that "the claim that there is evidence for the presence of a black hole event horizon on the basis of a lower quiescent luminosity for black holes than for neutron stars is unproven" (Jonker et al. 2007). While acknowledging that the work of Jonker and his colleagues on 1H1905+000 is important, Narayan and McClintock, along with other experts on X-ray binary systems, take exception to their strong conclusions. They point out that optical observations suggest an orbital period for 1H1905+000 of about one hour, whereas the orbital periods in his team's sample range from four to twelve hours. A shorter period means higher orbital speeds, which would tend to make it difficult for matter to flow onto the neutron star. Since no known black hole system has such a tight orbit as 1H1905+000, comparing the systems may be like comparing apples and oranges.

In a review article, Jean-Pierre Lasota of the Institut d'Astrophysique de Paris concedes that it is worthwhile to kick a paradigm every now and then to see if it falls over, but concludes, "I think the 1H1905+000 kick is too weak even to shake the faint black hole paradigm. But observers should of course keep trying"(2008).

Observers are also continuing to search for and study more stellar black holes in our galaxy and beyond. Their goals are to find out which fraction of stars become black holes, how massive they can get, how fast they can rotate, and how matter behaves as it approaches the event horizon.

It has been established with fairly high reliability that stellar black holes must have masses more than three times that of the Sun. What is not known is whether there is a limit to how massive stellar black holes can get. Computer simulations of the evolution of massive stars indicate that stellar black holes with a mixture of elements approximately equal to that of the Sun cannot get much bigger than about fifteen to twenty times the mass of the Sun. Stars destined to become black holes may be much more massive than that throughout most of their lives, but toward the end, violent pulsations eject most of the stars' outer envelope. When their central nuclear energy source finally plays out, the implosion of the central regions, or core, of the stars creates a black hole of five to twenty solar masses, and in the process sends a colossal shock wave racing outward that blows away what's left of the star in a supernova.

In 2007, Jerome Orosz of San Diego State University and his colleagues, who included Jeffrey McClintock and Ramesh Narayan, published a paper describing what they claimed was the most massive stellar black hole detected so far (Orosz et al. 2007). The black hole, known as M33 X-7 (figure 24), is in M33, a nearby—meaning only about 3 million light-years away—galaxy. X-ray observations with Chandra and optical observations with the Gemini telescope on Mauna Kea showed that the mass of the black hole is 15.7 times that of the Sun.

A key to understanding the nature of M33 X-7 is that its X-ray emission undergoes an eclipse every three and a half days. X-ray observations with Chandra and XMM-Newton revealed how long this black hole is eclipsed by its companion star, which indicates the size of the companion. Observations with the Gemini telescope tracked the orbital motion of the companion around the black hole, giving information about the mass of the two members of the binary. Other observed properties of the binary were also used to help constrain the mass estimates of both the black hole and its companion.

The details of how gas plunges from the accretion disk into the black hole depends on how rapidly the black hole is rotating. In general, gas can get closer to the event horizon of a rapidly rotating black hole than it can in a slowly rotating one. This means that the gas is hotter, and the X-rays are more energetic and more numerous on average. Analysis of the power radiated by the infalling matter in X-rays, and the distribution with energy, or spectrum, of these X-rays then gives a fairly precise estimate of the rate of rotation of the black hole, assuming that its spin is approximately aligned with the plane of the orbit of the two stars—usually a good assumption. This is essentially the same method astronomers have used for decades to measure the radii of stars. Observations of the luminosity, or energy output, of a star and its temperature can be used together with well-tested formulas for radiation from hot objects to compute the size of the object. This is called the continuum-fitting technique, since it uses the continuum, or smooth radiation from the accretion disk, to calculate the radius of the inner edge of the disk. In the case of an accretion disk around a black hole, the location of the disk's inner edge depends on the rotation rate, so the determination of its size can then be translated into a rotation rate.

Applying this technique, the M33 X-7 group found that the M33 X-7 black hole rotates at 84 percent of the maximum rate, the speed of light. Black holes have been characterized as the simplest objects in the universe because only two parameters, mass and spin, are necessary to characterize them completely. As a result, for M33 X-7, we have a complete description of an object 14.5 miles across, about the size of a large comet, that is 3 million light-years away.

On September 14, 2015, black hole research, and astronomy as a whole, entered a new era when the Benjamin Abbott of Caltech and 1,012 coauthors of the LIGO Scientific Collaboration and Virgo Collaboration reported that two detectors of the Laser Interferometer Gravitational-Wave Observatory (LIGO)—one in Hanford, Washington, and one in Livingston, Louisiana—simultaneously (taking into account a delay of seven milliseconds between the Louisiana and Washington detectors) had observed a transient gravitational-wave signal. The signal matches that predicted by general relativity for the inspiral and merger of a pair of black holes, and the ringdown of the resulting single black hole. The initial black hole masses are estimated to be thirty-six and twenty-nine solar masses. The final black hole mass is sixty-two solar masses, with three solar masses of energy being radiated away in gravitational waves. This black hole, named GW150914, is by far the most massive stellar-mass black hole known. On December 26, 2015, LIGO observed another gravitational wave signal produced by the coalescence of two black holes, but these had significantly lower masses, estimated to be 14.2 and 7.5 times as massive as the Sun.

X-RAY & OPTICAL

24 | M33 X-7 STELLAR BLACK HOLE

An artist's depiction of a stellar-mass black hole discovered in the nearby galaxy M33. The black hole, revolving around a star (the large blue object) about seventy times more massive than our Sun, is itself almost sixteen times the Sun's mass. Some material blowing away from the blue companion star is captured in a disk (shown in orange) around the black hole. As matter in the disk swirls toward the black hole, frictional forces heat it to millions of degrees, and it emits X-rays. The inset shows a composite of data from Chandra (blue) and the Hubble Space Telescope. The bright objects in the inset are massive young stars around M33 X-7, and the blue Chandra source is M33 X-7 itself.

9

We have demonstrated the existence of an astrophysically efficient mechanism for extracting rotational energy from a Kerr hole: electromagnetic braking resulting from fields supported by external currents.

Roger Blandford and Roman Znajek, "Electromagnetic Extraction of Energy from Kerr Black Holes," 1977

CYGNUS X-1, MICROQUASARS, AND THE GALACTIC JET SET

In 1971, a team of scientists at American Science & Engineering in Cambridge, Massachusetts, used the Uhuru X-ray satellite to discover that the X-ray emission from a bright X-ray source known as Cygnus X-1 is highly variable on a time scale of less than a second. Over the next few years, Cygnus X-1 would be studied intensely, not just with Uhuru but also with optical and radio telescopes and with X-ray detectors on rockets. A strong case was made that Cygnus X-1 is a black hole (such as the one seen in figure 20, pages 68–69). At that time, theorists had discussed black holes, but there was still considerable doubt as to whether they actually existed. It took a decade or more before the astronomical community—including the black hole guru, Stephen Hawking—accepted their existence.

Cygnus X-1, as the granddaddy of black holes, has not faded into the background with age. At a distance of 6,070 light-years from Earth, it is one of the brightest stellar black hole X-ray sources and remains the subject of intensive study as astrophysicists strive to learn more about the nature of black holes and how matter behaves when it encounters them.

The new distance determination, made in 2011, was part of an intensive campaign by scientists from six universities, including Jeffrey McClintock and Ramesh Narayan, to determine as much as possible about the binary system containing Cygnus X-1 and its companion star. What we now know is this: Cygnus X-1 has a mass 14.8 times the mass of the Sun, and its blue supergiant companion star has a mass of 19.2 solar masses. Cygnus X-1 and its companion orbit around a common center of gravity every 5.6 days. The distance between the two stars is only one-fifth the distance of Earth from the Sun. Gas is flowing away from the supergiant in a fast wind, and the black hole is focusing some of this gas into an accretion disk around the black hole. As the gas spirals into the black hole, it is whipped to high speeds by intense gravitational forces and heated to about 10 million degrees Celsius. This hot disk is one of the brightest X-ray sources in the sky.

Using the same technique as for M33 X-7, investigators found that Cygnus X-1 rotates at greater than 95 percent of the maximum rate. The event horizon is whirling around at more than eight hundred times a second. This is an extremely important result (although it needs confirmation). For one thing, it implies that an enormous of amount of energy is stored in the rotating black hole.

An independent estimate of the spin of a black hole can be obtained from X-ray fluorescence, a method dating back to the discovery of X-rays by Wilhelm Roentgen in 1895. X-ray fluorescence can occur when high-energy X-rays bombard a material, creating an unstable atom that gives off fluorescent X-rays. The phenomenon is widely applied in the investigation of metals, glass, ceramics, and building materials; for research in geochemistry, forensic science, and archaeology; and for analyzing old manuscripts and art objects such as paintings and murals. In astrophysics, the development of powerful X-ray telescopes such as Chandra and XMM-Newton has made it possible to use X-ray fluorescence to probe the behavior of matter near the event horizon of a black hole.

X-ray fluorescence happens when a high-energy X-ray knocks an electron free from the innermost energy level of an atom, creating an unstable atom (figure 25). An electron from an outer energy level immediately jumps into a lower-energy state, with the emission of an X-ray photon with a distinct energy specific to the particular atom involved. This is called an X-ray emission line because it produces a sharp spike or line in the observed distribution of X-rays with energy. Around a black hole, fluorescence occurs when high-energy X-rays produced by hot gas very near the black hole strike atoms in gas spiraling toward the black

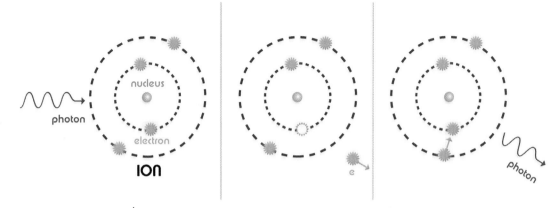

25 | X-RAY FLUORESCENCE

X-ray fluorescence from iron atoms, which occurs when a high-energy X-ray knocks an electron free from the innermost energy level of an iron atom, creating an unstable atom. An electron from an outer energy level immediately jumps into a lower-energy state, and an X-ray photon with a distinct energy specific to iron is emitted. Around a black hole, this happens when high-energy X-rays produced by hot gas very near the black hole hit iron atoms swirling around it, exciting the iron atoms to a higher-energy state. They almost immediately return to their lower-energy state with the emission of a lower-energy, fluorescent X-ray.

hole. Of particular interest is X-ray fluorescence from iron atoms, because they produce strong fluorescent lines at X-ray energies that X-ray telescopes can study.

An important feature of X-ray fluorescence, usually called X-ray reflection spectroscopy in astrophysics, is that the precise distance of the black hole system from Earth need not be known. This is especially valuable in that determining the distance to sources is often a thorny problem in astronomy. The Doppler effect and the effect of the strong gravity near the black hole change the shape of the spectral line, smearing it out to lower energies in a predictable way that depends on the rotation of the gas and the warping of space by the black hole. This in turn depends on how fast the black hole is rotating.

Applying the X-ray reflection method to Cygnus X-1 confirms that it is rotating at or near its extreme limit. This is an important point for a couple of reasons: It is one more argument that the event horizon exists, and it also gives us confidence in the X-ray reflection method, which is easier than the continuum-fitting technique to apply to supermassive black holes in distant galaxies.

The popular picture of a black hole is that of a cosmic whirlpool that pulls everything within its reach into a bottomless pit of space from which there is no escape. For most of the matter that drifts near a black hole, this is a fairly accurate picture. However, radio and X-ray telescopes have shown that there is more going on in the neighborhood of some black holes than a giant sucking sound.

One of the many peculiarities of Cygnus X-1 that puzzled researchers in the early years, before the launch of Uhuru, was the seemingly contradictory results of different experiments performed on short-duration rocket or balloon flights. Some reports said that Cygnus X-1 was a bright source radiating primarily low-energy X-rays. Other researchers disagreed, saying that the intensity was lower and that it radiated primarily higher-energy X-rays.

Another possibility was that everyone was right, that Cygnus X-1 is a highly variable source of X-rays. It was perfectly natural that varied short-term observations would yield different results. With an eye toward resolving this controversy, frequent observations of Cygnus X-1 were scheduled. These ultimately led, as discussed in the previous chapter, to the identification of Cygnus X-1 as a black hole. These observations also showed that Cygnus X-1 is indeed a highly variable X-ray source and that its variations are part of a general pattern that applies to most if not all black hole X-ray sources. These sources can exist in two broadly defined states: a soft, high state, and a hard, low one. These definitions use a terminology, common among physicists, wherein X-rays with higher energies are "harder," or more penetrating, than "softer," lower-energy X-rays. In the soft, high state, X-rays have an energy characteristic of a gas of about 10 million degrees Celsius and the source is bright, whereas in the hard, low state, the average energy of the X-rays is greater than in the soft state and the source is not as bright.

There is general agreement that these two states are a direct indication of what is happening around the black hole. In the soft, high state, the accretion disk extends inward, very close—if not all the way—to the event horizon. A hot cloud, or corona, surrounds the black hole. Radiation from the disk and the corona produces a bright X-ray source. In the hard, low state, in contrast, the accretion disk is truncated farther from the black hole. Some of the gas plunges into the black hole, while swirling magnetic fields sweep up the rest and hurl it away from the black hole at high speeds.

Jets of high-energy particles resembling awesome waterspouts are commonly observed blasting away from the vicinity of black holes at near the speed of light. These jets are seen around stellar-sized black holes in our galaxy and the supermassive black holes that lurk in the centers of galaxies. Stellar black hole jets extend for a light-year or more, whereas jets from supermassive black holes can reach across a million light-years.

How is it possible that a black hole, the ultimate sinkhole, can expel matter at such high speeds? The answer is that the hot gas, like almost all gas in the universe, is threaded by a magnetic field. As the gas swirls toward the black hole, the magnetic field is amplified by the compression and twisting motion of the matter. As has been known since the days of Michael Faraday, an English scientist who did pioneering investigations into the nature of electricity and magnetism almost two centuries ago, a swirling magnetic field can act like a generator, creating an electric field. The electric field then accelerates particles away from the black hole.

Roger Blandford and Roman Znajek, then at the University of Cambridge, first showed how this process could extract energy from a rotating black hole (figure 26). The spinning black hole twists the fabric of space around it, forcing the magnetic field in the infalling gas into a funnel shape, an electromagnetic tornado that flings magnetic fields and charged particles outward in two opposing jets. The more rapidly the black hole spins, the more efficient the process.

These ideas about the nature of jets need to be tested—a difficult proposition for objects millions of light-years away and processes that can take millions of years to unfold. That is why the discovery and subsequent study of the life cycle of jets from the black hole system XTE J1550 in our galaxy has stimulated such excitement among astrophysicists. XTE J1550, Cygnus X-1 (figure 27), and other black hole systems that exhibit powerful jets are called microquasars. This name emphasizes the sources' similarity to quasars, extraordinarily luminous sources produced by accretion into supermassive black holes in the centers of many galaxies.

XTE J1550 is a stellar black hole only seventeen thousand light-years away from Earth. At this relatively close distance, it has provided an excellent opportunity for astronomers, using Chandra and radio telescopes, to observe a black hole producing jets. Stephane Corbel of the University of Paris and his colleagues (Corbel et al. 2002) have tracked two opposing jets of high-energy particles emitted following an outburst from XTE J1550, first detected in 1998 by NASA's Rossi X-Ray Timing Explorer (RXTE), and saw that the entire process—from the expulsion of the jets to the slowing down and eventual fading of one jet—took only four years. The X-ray jets, which require a continuous source of trillion-volt electrons to remain bright, were first observed moving at about half the speed of light. Four years later, they are now more than three light-years apart and slowing down. One of the jets, as noted, has recently been observed to fade.

26 | JET OF HIGH-ENERGY PARTICLES

An artist's illustration of the formation of a jet of high-energy particles blasting away from a black hole in an accretion disk. The rotating black hole twists the fabric of space around it, forcing the magnetic field in the infalling gas into a funnel-like shape threaded by strong electric fields that accelerate charged particles outward in two opposing jets. The efficiency of the process increases as the black hole's spin speeds up.

As jets plow through the interstellar gas, the resistance of the gas slows them down, just as air resistance slows down moving objects on Earth. Although all jets are believed to decelerate in this way, observations of XTE J1550 mark the first time jets have been caught in the act of slowing down. The observed deceleration underscores the value of small stellar black holes in our galaxy for studying similar processes that occur in distant quasars and could take millions of years to unfold.

The black hole system SS 433 provides a different look at the ejection of matter in jets. A Chandra image of SS 433 shows two high-speed lobes of 50-million-degree Celsius gas, 3 trillion miles apart, on opposite sides of this binary black hole system. The binary system, which is between the two lobes, consists of a massive star and a black hole with a disk of hot matter. Material is ejected from this disk in narrow jets that slowly wobble or precess around a circle. This observation implies that the gas in the jets has been reheated, most likely by collisions between blobs of gas. Long-term optical monitoring observations have revealed that matter is ejected every few minutes from the vicinity of the black hole in bulletlike, gaseous blobs. The blobs apparently travel outward at about a quarter of the speed of light for several months without colliding until a faster blob rear-ends a slower one, precipitating a pileup that reheats the gas. A study by Herman Marshall of MIT and his colleagues (2013) used Chandra data to show that the jets originate five times closer to the black hole than was previously thought, and that the mass of the companion star is about sixteen times that of the Sun. SS 433 is similar to the XTE J1550 binary system in that they both involve black holes that produce high-speed jets of gas. However, there are significant differences between the two systems. Researchers have observed that the X-ray-emitting lobes in XTE J1550 are much farther away from the black hole than those in SS 433, and that the X-rays from the XTE J1550 lobes appear to be produced by a magnetized cloud of highly energetic electrons, not clouds of hot gas, as they are in SS 433. These differences indicate that the conversion of the rotational energy of a black hole into jets can operate in a variety of ways.

The most powerful known microquasar is a source known as S26, in NGC 7793, a galaxy 13 million light-years from Earth. Observations made with Chandra and the European Space Observatory's Very Large Telescope in Chile show that S26 has blown a huge bubble of hot gas into the galaxy's cooler interstellar gas. The bubble is a thousand light-years across. Its size implies that S26 is tens of times more powerful than other known microquasars. Thin beams or jets of fast-moving particles have slammed into the surrounding interstellar gas, creating a bubble of high-energy particles and hot gas, which is expanding at five hundred thousand miles per hour. The inflating bubble contains a mixture of hot gas and ultrafast particles at different temperatures.

Careful study of these black hole "jet sets" over the next few years may unlock more secrets of the enigmatic black holes, which turn out to be givers as well as takers in the cosmic scheme.

27 | CYGNUS X-1 WIDE-FIELD IMAGE

Image of a black hole (center) about fifteen times the mass of the Sun, in orbit with a massive blue companion star (right). Using optical observations of the companion star and its motion around the black hole, astronomers have made the most precise determination ever of Cygnus X-1's mass: 14.8 times the mass of the Sun. Probably it was almost as huge at birth because it has had comparatively little time to grow.

CHANDRA'S COSMOS

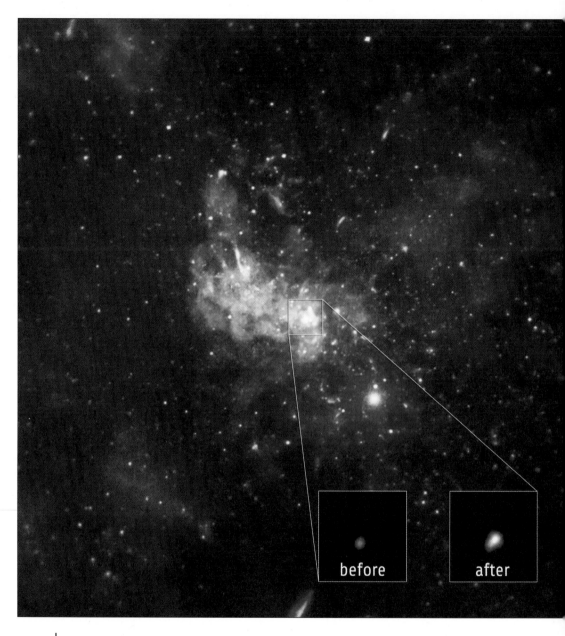

before

after

28 | SAGITTARIUS A*

A Chandra image of the region around Sagittarius A* (Sgr A*), the radio, infrared, and X-ray source at the center of the Milky Way. Low-, medium-, and high-energy X-rays are red, green, and blue, respectively. Sgr A* is in the white central area. The insets show Sgr A* before and after a large 2013 flare, when it showed a four-hundred-fold increase in brightness.

10

There is a way on high,

conspicuous in the clear heavens,

called the Milky Way, brilliant

with its own brightness.

Ovid, *Metamorphoses*, 8 CE

DOWNTOWN
MILKY WAY

E vidence for the existence of stellar mass black holes is very strong, as we have
seen. Optical and X-ray telescope observations of a couple of dozen binary
systems, each containing a normal star and an unseen companion, are very difficult to explain without invoking the presence of a black hole with a mass about
ten times that of the Sun. This finding is supported by theoretical calculations
that indicate that black holes are formed as a natural end state of the evolution of
very massive stars.

One of astrophysics' most significant accomplishments in the past twenty
years has been to establish that most, if not all, galaxies with masses greater than
about 10 billion Suns host supermassive black holes in their centers. This research
has been a broad-based effort, with radio, infrared, optical, and X-ray telescopes
all playing important roles. The resulting evidence is not as direct as it is for stellar mass black holes, but it is basically the same: The motions of stars and gas, in
the vicinity of the center of a galaxy, or energetic output from the central region,

require the presence of an unseen mass that is so great that there is no other explanation besides a black hole that fits with known physics. Chandra has been a major player in this work because intense X-radiation emanating from the center of a galaxy is one of the most reliable indicators of the presence of a supermassive black hole.

The nearest supermassive black hole to Earth is about twenty-six thousand light-years away, in the center of our home galaxy, the Milky Way. Before analyzing what we know about this supermassive neighbor, let us set the stage.

The word *galaxy* comes from the Greek word meaning "milky circle" or, more familiarly, "milky way." The band of white light across the night sky that we call the Milky Way was poetically described long before Galileo, or even Ovid, by astronomers as far back as Anaxagoras in the fifth century BCE. But it was not until Galileo trained his small telescope on the Milky Way that its nature was revealed. There he discovered a multitude of individual stars, "so numerous as almost to surpass belief."

Today we know that the Milky Way is a vast, rotating spiral of gas, dust, and hundreds of billions of stars. The Sun and its planetary system formed in the outer reaches of the Milky Way about 4.5 billion years ago. In the center of the Milky Way, also called the galaxy, is a bar-shaped galactic bulge that harbors a supermassive black hole (figure 29).

Cygnus X-3

Surrounding the central galactic bulge is a relatively thin disk of stars about two thousand light-years thick and roughly one hundred thousand light-years across. Giant clouds of dust and gas in the disk and the bulge absorb starlight and give the galaxy its patchy appearance.

The galaxy is home to generations of stars past. Many eventually become small, dense white dwarfs after passing through a bloated red giant phase. Other, more massive stars explode as supernovas, enriching the galaxy with heavy elements manufactured in their cores and leaving behind either neutron stars or black holes. The galaxy's bright stellar disk is embedded in a faint disk of old stars that is about three times thicker than the thin disk. Surrounding this disk is an extremely faint halo that contains both the oldest stars in the galaxy and dark matter, the dominant form of matter in the galaxy and in the universe as a whole.

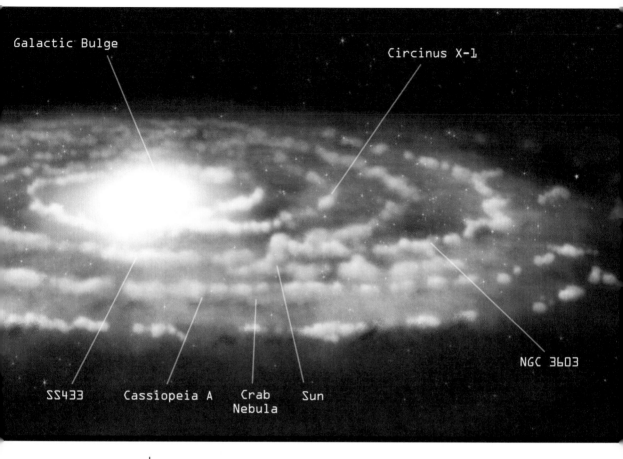

Galactic Bulge

Circinus X-1

SS433   Cassiopeia A   Crab Nebula   Sun   NGC 3603

29 | MILKY WAY

A schematic of our Milky Way galaxy depicting its central galactic bulge, its spiral arms, and the location of the Sun and selected X-ray sources.

It is estimated that the mass of dark matter in our galaxy is five times that of all the stars in the galaxy.

Most action in the galaxy takes place in its crowded center, the bustling "downtown," so to speak, of our galactic metropolis. Using Chandra, scientists are trying to determine how this relatively small patch of galactic real estate affects the evolution of the galaxy as a whole. For example, Chandra data indicate that large quantities of extremely hot gas are apparently escaping from the Galactic Center. This outflow of gas—enriched with elements such as iron, carbon, and

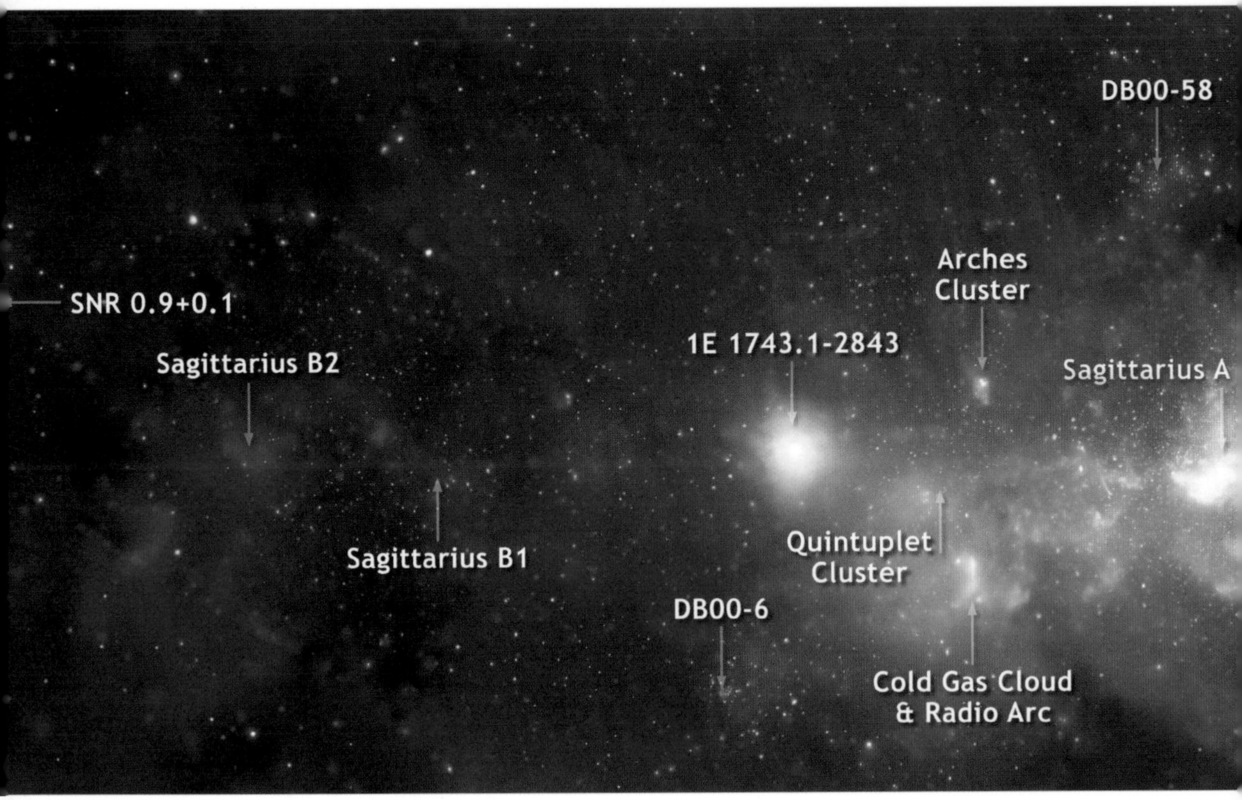

Labels on image:
DB00-58
Arches Cluster
SNR 0.9+0.1
1E 1743.1-2843
Sagittarius A
Sagittarius B2
Sagittarius B1
Quintuplet Cluster
DB00-6
Cold Gas Cloud & Radio Arc

30 | GALACTIC CENTER

A four-hundred-by-nine-hundred-light-year panorama of the central region of the galaxy. Sagittarius A*, a supermassive black hole twenty-six thousand light-years from Earth, is in the center of the galaxy, inside the central white patch labeled Sagittarius A. This photo combines Chandra images of low-energy (red), intermediate-energy (green), and high-energy (blue) X-rays.

silicon from the frequent destruction of stars—is distributed throughout the rest of the galaxy.

We live in the suburbs of the galaxy, far from the teeming, tumultuous Galactic Center. On the one hand, that's probably a good thing for fragile creatures such as ourselves. On the other hand, we are also incurably curious and would like to know what's happening there.

For optical telescopes that routinely look at objects billions of light-years away, examining a region only twenty-six thousand light-years distant shouldn't present

X-ray Thread

Sagittarius C

1E 1740.7-2942

DB01-42

much of a problem. However, it's not that simple. As in many major cities, air quality in the Galactic Center is terrible. Dust and gas produced by millions of massive stars make it impossible for even the most powerful optical telescopes to see into this region.

Fortunately, other options are available. Radio, infrared, X-ray, and gamma-ray radiation can travel through the galactic smog and be captured by telescopes sensitive to these forms of light. Using this information, astronomers have been gradually piecing together a picture of the center of the Milky Way. Chandra's unique ability to resolve X-ray sources as small as a tenth of a light-year across in the Galactic Center has led to major advances in our understanding of high-energy activity there. It has also exposed some mysteries.

A panoramic X-ray view extending four hundred by nine hundred light-years around the Galactic Center (figure 30) shows that even at distances of a few hundred light-years from the center of our galaxy, conditions are getting crowded and the energy level is increasing dramatically. Supernova remnants (SNR 0.9+0.1, probably the X-ray Thread, and Sagittarius A East), bright binary X-ray sources containing a black hole or a neutron star (called the 1E sources), and hundreds of unnamed pointlike sources that are neutron stars or white dwarfs light up the region. The massive stars in the Arches and other star clusters (sources with catalog numbers beginning with DB) will soon explode to produce more supernovas, neutron stars, and black holes. Infrared and radio telescopes have also revealed giant star-forming molecular clouds (Sagittarius A, B1, B2, and C, and the cold gas cloud near the Radio Arc), the edges of which are glowing with X-rays because of heating from nearby supernovas.

All this commotion takes place in a diffuse cloud of hot gas that shows up as an extended X-ray emission. The gas appears to have two components: a 100-million-degree Celsius part and a 10-million-degree part. This diffuse X-ray glow gets brighter toward the Galactic Center. The high temperature of the diffuse gas poses a problem—it should flow out of the Galactic Center in about ten thousand years, requiring continual replenishment and heating.

The stars in the Arches and other star clusters can supply the gas—the X-ray sources associated with these clusters show that they are blowing away a prodigious amount of matter—but they are unlikely to supply the heating mechanism. One possibility is that magnetic fields are involved in heating the gas or confining it to the center of the galaxy. The radio image, which is a tracer of magnetic fields in this region, shows that magnetic fields are certainly there. However, their structure does not appear capable of confining the hot gas.

Within a dozen light-years of the Galactic Center, the hurly-burly increases. Sagittarius A, the bright blob in the center, is composed of three main parts: Sagittarius A East, Sagittarius A West, and Sagittarius A*. Sagittarius A East is the remnant of a supernova that stirred things up about ten thousand years ago (plus, of course, the additional sixteen thousand years that it took the light to reach us from the Galactic Center).

Sagittarius A West is a spiral-shaped structure of gas that may be headed toward Sagittarius A*, which is located at the center of the galaxy. Sagittarius A*, or Sgr A* for short, is now synonymous with the Galactic Center (figure 28, page 90). Sgr A* gets its name from a radio source discovered by Bruce Balick and Robert Brown of the National Radio Astronomy Observatory, and it was so named because it is the brightest radio source in the constellation Sagittarius. The A* is from Brown, trained as an atomic physicist, who thought the newly discovered source was exciting, and excited states of atoms are often denoted with asterisks.

For the past twenty years, two groups—Andrea Ghez of the University of California, Los Angeles, and her colleagues, and Rainer Schödel and Reinhard Genzel of the Max Planck Institute for Astrophysics in Germany and their colleagues (Ghez et al. 2008; Schödel 2016)—have used infrared telescopes to track the motions of stars orbiting Sgr A*. They found that all the stars move around a common focal area containing a dark mass of 4.6 million Suns. Since the dark mass must lie inside the orbits of the stars, its radius must be less than 3.7 billion miles. This radius was refined using radio telescopes to a size of less than a few gravitational radii, smaller than the orbit of Mercury, for the black hole. The only plausible interpretation of this measurement is that that Sgr A* is a supermassive black hole. Ramesh Narayan and his colleagues at Harvard-Smithsonian have applied the arguments used to show that event horizons must exist in stellar black holes to show that an event horizon must exist for Sgr A* (Narayan and McClintock 2014).

One mystery involving Sgr A* is why it is not growing faster. All the matter being spewed out by those massive stars should provide the central black hole with a good, steady source of food, yet the X-ray power of Sgr A*, normally a good indicator of the rate of mass that is swallowed by a black hole, is unusually low. Failure to detect evidence of matter flowing into Sgr A* is not for want of looking, however. Chandra has been monitoring the source off and on for more than a decade; in 2012, an intensive five-week monitoring campaign was mounted, and another was launched in 2013. The bottom line is that less than 1 percent of the gas initially within Sgr A*'s gravitational grasp ever reaches the event horizon. Instead, most of the gas is ejected before it gets near the event horizon and has a chance to heat up. This makes Sgr A* a feeble X-ray source.

Explanations for the eating habits of Sgr A* abound. One is that the gas around it is simply too hot—blame the Sagittarius A East explosion for that. According to this idea, we are seeing Sgr A* in a quiet period, and it may get back on a regular feeding schedule in the future. Another is that the winds from the massive stars are blowing too fast to be captured by the supermassive black hole. A black hole is in a way like a big, slow dog. If a rabbit stays far enough away or moves quickly enough, it can escape, but if it ventures too close . . .

Chandra has caught Sgr A* in the act of snacking. By combining information from long monitoring campaigns by three X-ray telescopes, Chandra, XMM-Newton, and an X-ray telescope aboard NASA's Swift satellite, with observations by the Swift satellite itself, astronomers have been closely tracking the activity of the Milky Way's supermassive black hole over the past fifteen years. The study revealed that Sgr A* has been producing one bright X-ray flare about every ten days. However, in late 2013 and late 2014, the rate of bright flares from Sgr A* increased tenfold, to about one every day, and several large flares were observed. This increase happened about six months before a mysterious object called G2, a gas cloud possibly associated with a dust-enshrouded star, closely approached Sgr A*, and it happened again six months afterward. However, no significant increase in X-ray output from Sgr A* was observed just before, during, or after the passage of G2. More observations are needed to determine what caused the flaring: G2's passage; random fluctuations in the accretion process caused by tangled magnetic fields in infalling gas; or an asteroid-sized object falling into the black hole.

While the coincidence of G2's passage with the X-ray surge from Sgr A* is intriguing, astronomers see other black holes that seem to behave like Sgr A*. Therefore, it is possible that this increased chatter from Sgr A* may be a common trait among black holes and unrelated to G2. For example, the increased X-ray activity could be due to a slackening in the strength of winds from nearby massive stars that feed material to the black hole.

Even taking into account the large flares, the X-ray luminosity of Sgr A* is still puny when averaged over a week or so. There are indications, however, that this was not always the case. It is likely that Sgr A* had at least two major outbursts in the past few centuries. Evidence of the black hole's past activity has been found in the form of X-ray light echoes.

The X-ray echoes are produced by the X-ray fluorescence process described earlier in connection with stellar mass black holes. X-rays from an outburst generated by large clumps of matter falling into the black hole strike a cloud some distance from the black hole. Iron atoms in these clouds absorb these X-rays and emit them as fluorescent X-rays, generating the echo.

Several studies using Chandra, XMM-Newton, and Suzaku (an X-ray satellite developed jointly by the JAXA, the Japanese space agency, and NASA) in the past decade have uncovered evidence of X-ray echoes. Chandra's ability to precisely locate the origin of the reflected X-rays helps to pin down which clouds are reflecting the X-rays and how far they are from Sgr A*. Evidence of multiple outbursts was detected in clouds thirty to three hundred light-years away from Sgr A*. A likely possibility is that the observed flares were echoes of bright flares produced by major accretion events—a disrupted star or planet fell into the black hole. Some of the X-rays produced by these episodes traveled straight to Earth and arrived thirty to three hundred years ago. Other X-rays reflected off the gas clouds and arrived in time to be detected by Chandra.

The X-ray echoes enable astronomers to do "cold case" research on the past behavior of Sgr A*. They show that it was at least a million times brighter in the past few hundred years. These intense flares could have been produced by the partial disruption of a star or planet that came too close to Sgr A*, or maybe it did a little housecleaning and just swept up a bunch of gas and dust that been accumulating there over the years.

Astronomers are hopeful that such an event will occur again soon. Not to worry—even if it were to increase a millionfold, Sgr A* is so far away that it would show up as just another bright X-ray source such as Cygnus X-1.

31 | CYGNUS A OPTICAL IMAGE

An optical image of Cygnus A, 700 million light-years from Earth. It is the small, fuzzy, butterfly-shaped object at center.

11

We dance round in a circle and suppose,

But the Secret sits in the middle and knows.

Robert Frost, "The Secret Sits," 1942

THE SECRET
IN THE MIDDLE

The concept of an engulfing whirlpool appears in the literature of virtually every mythology and is a persistent theme in science fiction. Current astrophysical research aids and abets this concept, showing once again that the truth can be as strange as—or even stranger than—fiction.

Indeed, understanding supermassive black holes, how they are formed and grow, and how they affect their environment is a major quest for astrophysicists today. One of the primary ways to study black holes is through the radiation emitted by gas falling into them or by gas that, as often happens, is ejected from them.

To get a feeling for the importance of supermassive black holes in the cosmic scheme of things, consider the following: It has been estimated that approximately one quarter of the total cosmic radiation emitted since the Big Bang has come from matter swirling around and into supermassive black holes. A significant amount of this radiation is emitted in the X-ray band, so it is not surprising that Chandra's observations of supermassive black holes have played a key role in exploring these cosmic maelstroms and understanding their far-reaching effects.

The first known astrophysical black hole, Cygnus A (figures 31–32), may have been discovered more than seventy years ago, although the discoverers didn't truly recognize that they had found it, so maybe it doesn't count. Cygnus A—not be confused with Cygnus X-1, the stellar mass black hole that was the first black hole discovered—has been the focus of attention and controversy among astronomers since it was found many decades ago. The history of scientists' attempts to come to grips with what is happening in Cygnus A shows the value of observing a source with different types of telescopes, and it serves as a cautionary tale to all who would leap to conclusions on the basis of limited data.

In 1946, many engineers and scientists in the United Kingdom and Australia used talents and technology developed during World War II to explore the sky with radio telescopes. They were especially interested in a region of the constellation Cygnus that had been identified as a strong source of radio waves by astronomy pioneer Grote Reber, who had used a $2,000 homemade telescope in his back yard in Wheaton, Illinois, to make a radio map of the Milky Way.

In 1946, British physicists Stanley Hey and his colleagues S. J. Parson and J. W. Phillips used modified anti-aircraft radar antennae to study the Cygnus region in detail. They found a strong, rapidly fluctuating radio source that they named Cygnus A. They concluded: "It appears probable that such marked variations could only originate from a small number of discrete sources" (Hey, Parson, and Phillips 1946). The idea is that, if the radiation from a source is observed to vary over a certain period, say twenty seconds, then the size of the source must be less than the distance a shock wave or other disturbance that causes the variability can travel in twenty seconds. Otherwise, the variation would be weak and undetectable. The fastest speed that a disturbance can travel is the speed of light—299,792 kilometers (186,282 miles) per second. The maximum size of a source that varies in twenty seconds is thus about 20 × 300,000 = 6,000,000 kilometers—roughly the size of a large star with a diameter four times that of the Sun. This led some astronomers, most notably Martin Ryle, to propose that Cygnus A and similar sources were a new type of star that shone at radio wavelengths but was invisible at optical wavelengths. He called these objects radio stars.

By 1950, more detailed observations by Ryle and others in England and Australia showed that the radiation from Cygnus A was in fact quite steady; its apparent variability was due to the bending of radio waves by clouds of ionized gas in Earth's ionosphere. A similar effect causes stars to twinkle.

Paradoxically, even though the radio star hypothesis was invented to explain the variations of Cygnus A and other sources, the discovery that these variations were due to Earth's atmosphere and not intrinsic to Cygnus A did not cause the supporters of this hypothesis to abandon it. They pointed out, correctly, that stars twinkle and planets do not, because stars are pointlike and planets are disklike (so the twinkling is washed out for an Earthbound observer). Therefore, since Cygnus A twinkles, it must be pointlike, or at least it must have a small angular size.

At a conference in London in 1951, the radio star hypothesis was debated by some of the leading astronomers of the day. Ryle and George McVittie defended the radio star hypothesis, while Thomas Gold and Fred Hoyle argued that the sources of the radio emission were beyond the Milky Way galaxy. "Why . . . does not one find any identifiable visual object where those very near radio stars are supposed to be?" asked Gold. Gold went on to point out that the fifty or so radio sources known at that time were not concentrated, like the stars in our galaxy, but rather far-flung, like galaxies that are much more distant, and he proposed that the sources were radio galaxies. "It cannot be ruled out," he said, "that other galaxies may behave quite differently from our own, for it is known that there are very different types."

"It does not seem to me that an extragalactic nebula [meaning a galaxy] can do the trick," responded McVittie, who pointed out that very few of the galaxies were known to be radio emitters and that most of the radio stars could not be associated with a known galaxy.

"I think the theoreticians have misunderstood the experimental data," Ryle added, emphasizing along with McVittie that "there is as yet no evidence to suggest that other extragalactic nebulae emit radio waves having a much greater intensity than our own galaxy." He went on to make the important point that the normal mechanisms for producing radio waves, such as radiation from a gas, would not work very well for galaxies.

Hoyle, in response to Ryle's slam at the theoreticians, responded that "the boot is really on the other foot, for Professor McVittie and Mr. Ryle have dogmatically asserted that the discrete sources cannot be of extragalactic origin, although . . . five have been found to correspond to nearby extragalactic nebulae."

The whole argument was settled before the year was out. Francis Graham Smith used an improved radio telescope in Cambridge, England, to get a much more accurate position of Cygnus A. Smith airmailed his results at once to Walter Baade at the California Institute of Technology in Pasadena. Within a few weeks, Baade was in the observing cage of the two-hundred-inch telescope on Mount Palomar. He focused the powerful telescope on the position given by

Smith and took two photographs, one in blue and one in yellow light. The next afternoon, he developed the photographs.

"I knew something was unusual the moment I examined the negatives," Baade recalled. "There were galaxies all over the plate, more than two hundred of them, and the brightest was at the center. It showed signs of tidal distortion, gravitational pull between the two nuclei—I had never seen anything like it before. It was so much on my mind that while I was driving home for supper, I had to stop the car and think" (see Pfeiffer 1956; Robinson et al. 1965).

Baade concluded that he was seeing two galaxies in collision. This was especially exciting to him because he and Lyman Spitzer Jr. had published a paper earlier in the year in which they discussed some observable effects of such cosmic train wrecks. Baade's colleague Rudolph Minkowski noted the coincidence: "Baade and Spitzer invented the collision theory," he is reported to have said, "and now Baade finds evidence for it in Cygnus A."

Baade, angered by Minkowski's remark, challenged him to a bet of a thousand dollars that Cygnus A is the result of a collision. They settled instead on a bottle of whiskey and on the acceptable proof: emission lines of high excitation. Such lines are produced by atoms and ions in a low-density gas that has a temperature of ten thousand degrees Celsius or more. The spectra of stars, in contrast, show predominantly absorption lines due to relatively cool material in the atmosphere of the star. The basic idea is that a collision should produce a cloud of hot, low-density gas that could explain the emission lines in the spectrum of Cygnus A.

Within weeks, Minkowski had taken the spectrum of Cygnus A with the Palomar telescope and conceded the bet, although the payoff was not exactly what Baade had anticipated. "For me, a bottle is a quart," Baade said, "but what Minkowski brought was a hip flask. . . . Two days later, it was a Monday, Minkowski visited me in order to show me something—he saw the flask and emptied it" (quoted in Robinson et al. 1965).

The ironic part of this story is that a dozen years later, most experts would agree that Minkowski might have been justified in drinking the whiskey. A series of startling new discoveries had cast doubt on the idea that the spectrum of Cygnus A is the site of colliding galaxies. The spectrum of Cygnus A showed that the emission lines were all redshifted by the same amount. This in itself was not unexpected, since two decades earlier Edwin Hubble and his colleagues had shown that the spectra of distant galaxies show a redshift proportional to their distance. The difficulty was that the measured redshift implied that Cygnus A was very distant—possibly a billion light-years distant.

At such a large distance, Cygnus A would need an extremely powerful energy source to produce the observed radio and optical radiation. It seemed impossible that galaxies in collision could generate such intense power. This problem was made even worse in the mid-1950s by additional radio observations and theoretical work. The conclusion of these efforts was that radio waves from Cygnus A and other radio-emitting galaxies are produced by high-energy electrons spiraling around magnetic field lines. This process, which was first observed in a particle accelerator called a synchrotron, is called synchrotron radiation.

At a meeting of radio astronomers in Paris in the summer of 1958, Geoffrey Burbidge discussed the implications of synchrotron radiation coming from Cygnus A. He showed that the energy needed to produce the high-energy particles was much greater than the expected energy from a collision of galaxies.

Then still more problems for the collision theory became apparent when R. C. Jennison reported on new measurements of the structure of the Cygnus A radio source. It has the shape of a great cosmic dumbbell, with two huge lobes of high-energy particles located more than five hundred thousand light-years apart. The galaxy, which is several times smaller than the lobes, is located in the middle. It was beginning to look as if the radiation from Cygnus A was not due to galaxies in collision but to some mysterious explosive process in which high-energy particles were being blown out of the galaxy.

The search for a way to produce high-energy particles soon took an unexpected turn. Fred Hoyle pointed out that if other powerful radio galaxies had sources as large as Cygnus A's, they could be seen at distances much greater than that of Cygnus A. If the sources were all about the same intrinsic size, then a measurement of their apparent size could give astronomers vital information about the nature of the universe. For example, the Steady State universe, a concept championed by Hoyle, Thomas Gold, and Hermann Bondi, would show a steady decrease in angular size, whereas the Big Bang universe would show a decrease only up to a point before it started to increase because of the curvature of space.

This insight spurred a flurry of activity and intense work by radio astronomers to determine the size of faint radio sources and settle the issue of whether the universe was in a steady state or had originated in a Big Bang. They found many new radio galaxies, and scientists on various sides of the issue soon became embroiled in a bitter debate on the significance of the results.

In the meantime, radio astronomers found that some of the newly discovered radio sources were difficult or impossible to resolve optically, even with their largest and most sensitive telescopes. Burbidge and his colleagues, including his wife, Margaret, and Caltech astronomer Allan Sandage, began to look at the nuclei of galaxies, where they suspected the necessary source of energy might be found.

Thomas Matthews and Sandage used the Palomar two-hundred-inch telescope to identify one of the radio sources, called 3C48, with a faint blue starlike object. The object presented a profound puzzle. It had the colors of a white dwarf, but it also showed a wisp of nebulosity that had never been seen in association with a white dwarf. Most mystifying of all was its spectrum. Like Cygnus A's, it contained emission lines, indicating gas at ten thousand or more degrees Celsius in the vicinity of the object. Unlike with Cygnus A, or any other object they had ever encountered, they could not match these emission lines with any known elements.

Over the next two years, many spectra of 3C48 were taken by a number of astronomers, but no one could identify the emission lines. Then, in 1963, the Australian radio astronomers Cyril Hazard, M. B. Mackey, and A. J. Shimmins achieved a breakthrough by using the fact that the Moon was eclipsing another radio source, 3C273. Since the position of the edge of the Moon is known very accurately for any time, by carefully timing the disappearance and reappearance of the source, they could measure the position and size much more precisely than was previously possible. They found that 3C273 is a double radio source, with one component located on top of a blue starlike object.

With this accurate position, Maarten Schmidt of Caltech used the Palomar two-hundred-inch telescope to photograph 3C273 and take a spectrum. The photograph showed a starlike object with a faint wisp or jet off to one side. The spectrum showed mysterious, broad emission lines like those seen in the spectrum of 3C48.

Then Schmidt had one of those simple yet profound insights that seem so obvious in retrospect. He showed that the mysterious lines were due to well-known transitions between energy states in a hydrogen atom. They were merely shifted to the longer wavelengths (to the red end of the spectrum) by about 16 percent.

This was not the largest redshift observed to that point—Minkowski had shown that the galaxy 3C295 and its associated galaxy cluster have a redshift of 46 percent. The problem, as expressed by Schmidt in his 1963 paper describing his discovery, was "the unprecedented identification of the spectrum of an apparently stellar object in terms of a large red-shift." How could a starlike object have such a large redshift?

Schmidt offered two explanations: 3C273 is (1) a stellar object in which the redshifts are caused by a strong gravitational field, an idea difficult to reconcile with the details of the spectrum; or (2) the nucleus of a galaxy that is moving away from us at a velocity of 106 million miles per hour.

Using the Hubble redshift-distance relation, the latter interpretation implies that 3C273 is more than 2 billion light-years away. This in turn means that the nucleus of this galaxy is about one hundred times brighter than the luminosity of an entire bright galaxy. Yale University astronomers Harlan Smith and Dorrit Hoffleit used Harvard Observatory's historical photographic plate collection to show that 3C273 had been varying its light output significantly over the period of a year.

Soon after Schmidt's discovery, Jesse Greenstein and Thomas Matthews at Caltech reexamined the spectrum of 3C48. Unlike with 3C273, hydrogen lines were not apparent, which explains why they may have missed the redshift interpretation. Nevertheless, by using the redshift hypothesis, they were able to fit the spectrum to lines from the elements magnesium, neon, and oxygen, all redshifted by 37 percent.

These discoveries happened so quickly that they were reported in the same issue (March 16, 1963) of the journal *Nature*. This date marks astronomers' discovery of one of the most important constituents of the universe: quasi-stellar radio sources, a.k.a. quasars, a.k.a. quasi-stellar objects, a.k.a. QSOs.

Is a quasar an extreme example of an explosive galaxy like Cygnus A, or is it a totally different object? What might be the source of the immense power in quasars and in the centers of radio galaxies? These questions were on the minds of astronomers and astrophysicists as they gathered in Dallas, Texas, in December 1963 at a symposium, "Quasi-Stellar Sources and Gravitational Collapse."

A smorgasbord of ideas was advanced for the energy supply of quasars: matter-antimatter annihilation, the chain reaction of supernovas, the collapse of a cluster of stars, and the collapse of a superstar millions of times more massive than the Sun. Most of the talk turned to the collapse of something, because, as Philip Morrison of MIT said, "The evidence strongly favors the look of a gravitational event" (quoted in Robinson et al. 1965).

The supermassive star model attracted the most attention, partly because it had been worked out in more detail, and partly because its authors were Hoyle and his colleague William Fowler of Caltech, two of the most innovative and respected astrophysicists of the day. However, their reputation did not immunize them from criticism. Freeman Dyson of the Institute of Advanced Study put his finger on a major problem: Once the supermassive star's collapse starts, it will be over in a day. Ways out of this dilemma, such as rapid rotation or fragmentation, were discussed, and there was mention of the ultimate fate of matter, of gravitational event horizons and other bizarre ideas that would eventually come to occupy center stage. These ideas involved black holes, but they were not

32 | CYGNUS A COMPOSITE IMAGE

A composite Chandra X-ray image (blue) and Very Large Array radio image (red) of Cygnus A showing a large, football-shaped cloud of hot gas extending over several hundred thousand light-years. The faint regions, or cavities, are evidence of repeated explosions from the supermassive black hole that is the source of Cygnus A's energy output. Visible light data (yellow) from Hubble and the Sloan Digitized Sky Survey are also seen here.

mentioned as such, because the term *black hole* would not be coined until 1967. In the meantime, the consensus was that Minkowski had been right to drink Baade's bottle of whiskey. Cygnus A most likely was not due to the collision of galaxies.

One suggestion, made by Edward Teller, was that quasars' prodigious energy source is the collision of huge clouds containing 10 million solar masses of matter and antimatter. Yet Teller's idea is not supported. Small bits of antimatter can be produced by collisions between particles with high energy, but there is no evidence for such large concentrations of antimatter in the universe. Furthermore, the primary result of such collisions is gamma rays, not the combination of high-energy jets, magnetic fields, and radio, optical, ultraviolet, and X- and gamma radiation seen in quasars.

Nuclear power also seemed an unlikely source. Although that is what powers stars, it would be impossible to pack 100 billion Suns into a region one light-year in diameter. Stars more massive than the Sun radiate more efficiently, but such a dense collection of stars would collide and explode as supernovas or merge to form a supermassive star, which would then implode. Ultimately, scientists concluded, quasars' source of energy had to be gravity, the familiar force that has been known since Isaac Newton to operate throughout the universe.

But this was not Newton's brand of gravity—it was Einstein's, in which gravity is described not as a force but as a warping of space due to the presence of matter, and energy is the equivalent of matter. So the more energy you pack into a small space, the more mass that is concentrated there, and the greater is the warp in space and the stronger the concentration of energy, leading to a vicious cycle of unending collapse into . . . into what?

In the 1960s, research on the ultimate fate of massive stars was moving on a parallel track with work on quasars and radio galaxies. An increasing number of physicists were coming around to the viewpoint that the collapse of the core of a sufficiently massive star produces what John Wheeler of Princeton called a "black hole."

In 1964, Edwin Salpeter at Cornell University and Yakov Zeldovich of the Institute of Applied Mathematics in Moscow independently proposed that a stream of gas falling toward what they called "collapsed matter" or "collapsed objects" could in principle be heated to very high temperatures, where it would produce X-rays. Given enough matter, the X-rays would be detectable. Zeldovich and his colleague Igor Novikov proposed that a black hole that is part of a double star system would have the best chance of being detected.

The confirmation of this idea came in 1971 when combined X-ray, optical, and radio observations provided strong evidence that Cygnus X-1 is a black hole accreting matter from a bright blue companion star.

This discovery opened the way for the eventual acceptance of a black hole model as the central powerhouse of quasars. Donald Lynden-Bell and Martin Rees of Cambridge University showed that a gigantic black hole in the center of a galaxy could produce the necessary energy if it swallowed matter, or gas, in an amount equal to about one Sun per year. The gas would form an accretion disk around the black hole. Because of friction between the gas particles in the disk, they would gradually lose energy and move inward or, in the words of Lynden-Bell and Rees (1971), "slowly run down into the central black hole just as water runs out of a bath." The glow from the outer portion of the heated disk would produce the optical radiation seen in quasars. X-radiation would be produced in the extremely hot inner parts of the disk.

In subsequent work, Lynden-Bell, Rees, and their colleague Roger Blandford showed that the swirling motion of the gas in the quasar accretion disk would strengthen any magnetic field present, creating two tornados of spinning magnetic field lines on the top and bottom of the disk. These whirling magnetic fields could then generate the powerful jets of high-energy particles and magnetic fields that are seen blasting out from the centers of radio galaxies.

As telescopes became more sensitive and better able to probe the inner parts of radio galaxies and study quasars in more detail, the supermassive black hole model gained acceptance from virtually all astrophysicists. Radio, infrared, and optical telescopes have detected matter swirling around dark, massive objects in the centers of more than two dozen galaxies, and bright central pinpoints of X-radiation have been observed in most of these objects. Radio and X-ray jets have been traced back to the nuclei as well, with Cygnus A providing one of the most spectacular examples (figure 33; see also figure 22, page 73, the Pictor A galaxy).

In a sort of epilogue to the bet between Baade and Minkowski, the question of whether Cygnus A is due to a collision of galaxies is still unresolved. The unusual shape of the galaxy is due to dust and gas clouds, but it may harbor a double or even triple nucleus. Its energy is very likely produced by the accretion of gas into a gigantic black hole, but the origin of this gas may have been a collision with another galaxy, because Cygnus A is sitting in a cluster of galaxies. If they were still around, Baade and Minkowski would probably have to split a bottle of whiskey and call it a draw.

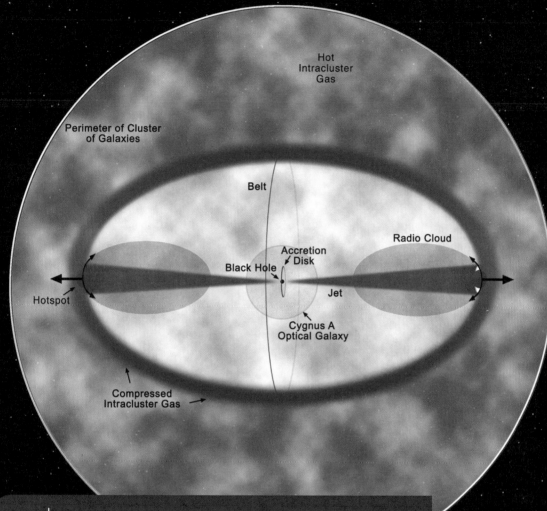

Hot
Intracluster
Gas

Perimeter of Cluster
of Galaxies

Belt

Radio Cloud

Accretion
Disk

Black Hole

Jet

Hotspot

Cygnus A
Optical Galaxy

Compressed
Intracluster Gas

33 | CYGNUS A SCHEMATIC

An outline of a model for the Cygnus A system. It consists of a supermassive black hole
embedded in a large galaxy, which is in turn embedded in a cluster of galaxies with a huge
reservoir of multimillion-degree Celsius intergalactic gas. Gas in the galaxy's central
regions forms an accretion disk around the supermassive black hole. As magnetized gas in
this disk spirals toward the black hole, its rotation generates a pair of powerful jets of
high-energy particles that produce strong radio and X-ray emission and inflate a giant
cavity in the intergalactic gas. The bright belts may be due to inflation of the cavity in the
galaxy's equatorial plane. The intergalactic gas has a very low density but provides enough
resistance to slow the jets, which terminate in radio- and X-ray-emitting hotspots about
three hundred thousand light-years from the central black hole.

34 | CHANDRA DEEP FIELD SOUTH

An X-ray full-field image of Chandra Deep Field South, in the constellation of Fornax. This four-million-second exposure made by Chandra was, until recently, the deepest X-ray image ever obtained of the sky (the exposure in the same field has now reached 7 million seconds, or about seventy days). Most X-ray sources seen in this image—some of which existed when the universe was less than 950 million years old—contain rapidly growing supermassive black holes.

12

If it looks like a duck, and

quacks like a duck, we have at least to

consider the possibility that we have

a small aquatic bird of the

family Anatidae on our hands.

Douglas Adams, *Dirk Gently's Holistic Detective Agency*, 1987

DUCKS
UNLIMITED

While Walter Baade and Rudolph Minkowski were gathering data on Cygnus A, their colleague Carl Seyfert was using the hundred-inch telescope on Mount Wilson to make detailed studies of the central regions of six spiral galaxies. These six galaxies had evidence of unusual activity in their nuclei. In contrast to the spectra of normal galaxies, the nuclei of these galaxies were characteristic of hot ionized gas clouds. Seyfert's observations showed that the hot gas clouds were moving rapidly out of the galaxies' nuclei at speeds ranging from several hundred thousand miles per hour up to millions of miles per hour (see Seyfert 1943).

This startling result was ignored for twenty years when Margaret and Geoffrey Burbidge, along with Kevin Prendergast, published an article discussing what they called the "Seyfert galaxies," a name that stuck (Burbidge, Burbidge, and Prendergast 1963). They emphasized that gas must flow out of the nuclear regions of a Seyfert galaxy at high speeds. Over the next few years, the Burbidges continued to tell anyone who would listen that some very interesting things were happening in the nuclei of galaxies, and in 1963, they published, in collaboration with Allan Sandage, an extensive review article entitled "Evidence for the Occurrence of Violent Events in the Nuclei of Galaxies," which should have jump-started the research on Seyfert galaxies. However, by the time the article was published, Maarten Schmidt's paper "3C273: A Star-Like Object with Large Red-Shift" had appeared in *Nature*, and the race was on to discover and explain quasars.

In the excitement that followed, it took a while for astronomers to understand that the engine powering both quasars and Seyfert galaxies is a supermassive black hole located in the nucleus of a galaxy. The nuclei of galaxies in which supermassive black holes are actively accreting gas and producing powerful radiation are called active galactic nuclei, or AGNs. A quasar is an extreme example of an AGN in which the active nucleus outshines the entire galaxy.

Differences among various AGNs are thought to owe to a variety of factors that relate to both the evolutionary stage of the AGN and the viewing geometry. These include the size of the central supermassive black hole, the amount of gas available to feed it, and whether the view of the black hole is obscured by a disk or torus of gas and dust around it. Type 1 AGNs show evidence of energetic activity in the form of hot, rapidly moving gas clouds, bright radio emissions, and low-energy X-ray emissions. Type 2 AGNs show evidence of less energetic activity in the form of scattered light and gas moving at moderate speeds, some radio activity in the form of jets, and no low-energy X-rays.

The AGN unified model, developed by Robert Antonucci of the University of California, Santa Barbara, and Joseph Miller of the University of California, Santa Cruz, building on the work of many other astronomers, proposed that Type 1 and Type 2 AGNs could have a common explanation—a donut-shaped veil of gas and dust around a central black hole. According to the unified model of AGNs, what is observed all depends on your point of view. The source looks different depending on whether it is observed from the top (through the hole of the donut; Type 1, in figure 35) or the side (the edge; Type 2, seen in figure 36).

The unified model explains many of the observations of active galaxies and has won widespread acceptance among astrophysicists. It turns out that Type 2 AGNs are necessary to explain X-ray background radiation. In 1962, Riccardo Giacconi and his colleagues on the historic rocket flight that discovered the first extrasolar X-ray sources also discovered that the X-ray sky is not dark. Rather, it is bathed in a uniform, diffuse glow called the X-ray background. A few years later, when quasars were discovered, some scientists suspected that the X-ray background might be caused by extremely distant AGNs.

This was a plausible idea: The numbers of AGNs and their power seemed to work out about right. But there was a problem. The spectrum of AGNs does not match the spectrum of the background X-rays. The ratio of high-energy to low-energy X-rays observed for X-ray background radiation is about four times greater than AGNs produce. You might say that the sources needed to explain the background radiation look like ducks and walk like ducks, but they do not quack like ducks. X-ray astronomers called this problem the "spectral paradox." One possible solution to the spectral paradox was that something other than AGNs are the source of the X-ray background.

Another possible solution was that some AGNs might be Type 2 quasars. In this case, they would show an excess of high-energy X-rays, not because the AGNs produce such an excess, but because donuts of gas around the supermassive black holes filter out low-energy X-rays. If many such sources were found with different-sized donuts and at different distances, the X-ray background would be explained. In other words, if the ducks have sacks over their heads, they might not sound like ducks, but they really are ducks.

Deep sky surveys with Japan's Advanced Satellite for Cosmology and Astrophysics (ASCA) and the Italian-Dutch Beppo SAX X-ray satellite gave support to the idea that other sources besides Type 1 AGNs are needed to produce the X-ray background glow. This research showed that about 30 percent of the X-ray background could be resolved into distinct sources. Given the limited angular resolution of these satellites, nothing more could be said. It would take Chandra to make further progress.

Soon after Chandra's launch, its greater sensitivity and focusing ability were put to work to solve this problem. This project, called Chandra Deep Field South (figure 34, page 112), was initiated by Riccardo Giacconi and continues under the leadership of Neil Brandt of Penn State University. From time to time, Chandra returns to observe a specific point in the southern constellation Fornax—an angular area about half that subtended the full moon. To date, Chandra has accumulated observing time of about seventy days. So far about a thousand X-ray sources have been detected in Chandra Deep Field South (Xue et al. 2012). An estimated 70 percent of these sources contain supermassive black holes. The faintest sources detected are 10 billion times fainter than the first extrasolar X-ray source observed by Giacconi and his team on the rocket flight in 1962, an improvement in sensitivity comparable to that achieved in going from naked-eye observations to the Hubble Space Telescope.

Chandra Deep Field South has been used to make a complete census of Type 2 AGNs. Chandra observations have been supplemented by observations

35 | BLACK HOLE WITH ACCRETION DISK AND TORUS

Top view. This artist's conception shows the constituent parts of a black hole system. In the center is the black hole itself, surrounded by a disk of hot gas (blue and green). A large donut, or torus, of cooler gas and dust (yellow and red) surrounds the system.

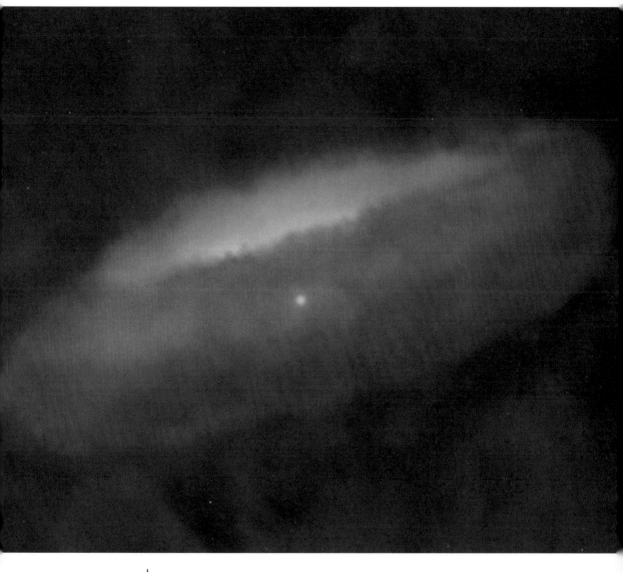

36 | BLACK HOLE WITH ACCRETION DISK AND TORUS

Side view. This artist's conception shows the same black hole and its surrounding disk of hot gas and torus of cooler gas and dust. The image suggests why X-rays are essential to exploring the nature of black hole systems: unlike optical emissions, they can penetrate the torus and thereby reveal information about the hidden black hole.

with the Hubble Space Telescope, Spitzer, and the Very Large Telescope in what is called the Great Observatories Origin Deep Survey South (GOODS-S) project. Many Type 2 AGNs are expected to be obscured so much that optical surveys would miss them completely. In principle, they can be detected by infrared telescopes, but infrared telescopes also detect dusty galaxies. At the expected distances of billions of light-years, it is impossible to distinguish between a shrouded AGN and a dusty galaxy with infrared observations. In contrast, obscured supermassive black holes are strong X-ray emitters, whereas the dusty interstellar medium in a galaxy is not, so Chandra can pick out Type 2 AGNs from the crowd. Furthermore, whereas results from previous X-ray telescopes were ambiguous because many different sources were in the field of view at very faint fluxes, Chandra's high-precision mirrors make it possible to pick out the sources without ambiguity. This procedure has increased the known number density of AGNs by a whopping 40 percent.

About a third of the sources detected are Type 1 AGNs; that is, they show no evidence of a dense, blanketing cloud of gas and dust. The remaining two-thirds of the sources appear to be veiled giant black holes, with a wide range of absorption. The conclusion is that the giant black hole population in the universe is three times as great as observations with optical telescopes indicate. It is as if previous bird counts had missed two-thirds of the ducks because they didn't look like ducks.

As Chandra observes more and more galaxies, evidence builds to support the idea that the number of powerfully radiating black holes is even greater than the estimate deduced from the X-ray background. This could be the case if a significant population of heavily obscured supermassive black holes exists, and several groups of scientists have been using Chandra to search for them. Investigators are making direct observations of specific sources and combing the Chandra archives in inventive ways—for example, they add up the observations of dozens of sources to see if there is a hint in the data of a heavily obscured AGN.

The stakes are high. In the words of one of the researchers, Andy Fabian of the University of Cambridge (2001), these results could "change our perspective on what the main sources of power in the universe are." Indeed, Fabian speculated that the increased numbers of known supermassive black holes already means that the total energy emitted by gas as it falls into them can be upgraded from one-quarter to as much as one-half of all the energy emitted by all the stars in the history of the universe.

37 | CHANDRA SURVEY IN BOÖTES CONSTELLATION

In contrast to Chandra Deep Field South, the Boötes field is a wide-field panorama covering an area forty times larger than the full Moon in the night sky and involving 126 separate Chandra exposures of five thousand seconds each. It is the largest contiguous field obtained by Chandra. Here red represents low-energy X-rays, green shows the medium range, and blue higher-energy X-rays. The red sources are mostly unobscured supermassive black holes, or Type 1 AGNs; green and blue ones are mostly obscured supermassive black holes, or Type 2 AGNs. The survey, which also used optical and infrared data, has identified more than six hundred obscured and seven hundred unobscured black holes about 6 to 11 billion light-years from Earth.

SHOWN FOR SCALE

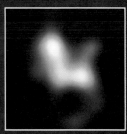

38 | NGC 3393 GALAXY

A spiral galaxy 160 million light-years from Earth. X-ray data from Chandra (blue) and optical data from Hubble (yellow) are combined. In the main image, the diffuse blue color represents hot gas near the galaxy's center and shows low-energy X-rays. The inset box, which depicts NGC 3393's central region as observed by Chandra alone, shows only high-energy X-rays, including emission from iron—a characteristic feature of growing black holes obscured by dust and gas. In the inset, two peaks of X-ray emission (at roughly 11 o'clock and 4 o'clock) are seen. These are actively growing black holes that generate X-ray emission as gas falls toward them and gets hotter. The obscured regions around the black holes block the large amounts of optical and ultraviolet light produced by infalling material.

13

Every single time you make a merger,
somebody is losing his identity.
And saying something different
is just rubbish.

Carlos Ghosn, CEO of Renault and Nissan, 2006

THE ORIGIN AND GROWTH OF SUPERMASSIVE BLACK HOLES

The recent discovery of dozens—about forty and growing as of 2016—of extremely distant supermassive black holes with masses of several billion Suns shows that enormous black holes were already in place less than a billion years after the Big Bang. Where did they come from?

One theory is that they grew from stellar black holes that formed from the first generation of stars about 10 million years after the Big Bang. These black holes would have had masses of about a hundred Suns, and they would have had to consume the surrounding gas at a prodigious rate—billions of Suns' worth—over the next billion years in order to explain the young supermassive behemoths. In principle this is possible, but recent computer simulations that take into account the back pressure of the radiation produced by the formation process indicate that growth of stellar mass black holes is unlikely to explain the existence of huge supermassive black holes in a billion years.

An alternative solution proposes to jumpstart the process with black hole seeds with masses of about a hundred thousand Suns, which formed from the collapse of either a supermassive star or a dense cluster of stars. Once formed, these black holes grow by accreting surrounding gas. Such objects are predicted to radiate mainly in the infrared and X-ray bands. Italian astrophysicist Fabio Pacucci and his colleagues recently performed a computer simulation of the formation and evolution of massive black hole seeds (Pacucci et al. 2015) and concluded that they may even be present as some of the faintest sources in Chandra Deep Field South. In a subsequent paper, Pacucci's group identified two objects, GOODS-S 29323 and GOODS-S 33160, on the basis of their infrared and X-ray properties, as direct collapse black hole candidates. The black hole masses are estimated to be between ten thousand and one hundred thousand times the mass of the Sun. The distances of these objects from Earth are estimated to be greater than 12.7 billion light-years, meaning that they are observed as they were less than a billion years after the Big Bang. Another similarly distant object, CR7, has also been put forth as a direct collapse black hole by a number of authors. CR7 is a very strong ultraviolet source but has not been detected in X-rays. All these sources will be the subject of intense investigation in future observing programs.

Another approach to understanding the precursors of the first supermassive black holes is to look around for galaxies that might contain prototypes of massive black hole seeds. Dwarf galaxies are the best bet for this search. In terms of being recognized by astronomers, they're the new kids on the block. It's not that dwarf galaxies haven't existed for billions of years, like normal-sized galaxies. They have. It's just that no one realized they existed until the 1930s. At that time, Fritz Zwicky (who discovered dark matter and first proposed that neutron stars exist) was embroiled in one of his many disputes with the great Edwin Hubble. Hubble was then at the peak of his fame and power at Caltech. He had proved that spiral nebulas are in fact galaxies similar to the Milky Way and had followed up this finding with the discovery that the universe is expanding.

Hubble maintained that all galaxies have roughly the same number of stars and the same brightness. Much smaller galaxies simply did not exist. Zwicky disagreed. He gave two arguments for the existence of dwarf galaxies. The first, based on the principle of the inexhaustibility of nature, reasons that nature is unlikely to allow only one type of galaxy to exist. Galaxies must come in all shapes and sizes, including dwarfs. Zwicky's second argument for the existence of dwarf galaxies was that if Hubble said that dwarf galaxies did not exist, then they almost certainly must exist. This was what Zwicky called "the method of negation and subsequent

CHANDRA'S COSMOS

construction" (1971), by which he meant one should look for statements, theories, or systems of thought that pretend to absolute truth and deny them. You are almost certain to be correct in doing this, Zwicky maintained, because it is extremely unlikely that anyone knows the absolute truth about anything.

In keeping with the second principle, Zwicky proposed to be constructive and search for dwarf galaxies. But Hubble and other authorities at Caltech refused to make available valuable time on the large telescopes to look for something that, they maintained, does not exist.

Undaunted, Zwicky acquired funds to build his own telescope, a small, eighteen-inch reflecting telescope especially designed to search large areas of the sky. Although this telescope had only one-thirtieth of the light-gathering power of the hundred-inch telescope on Mount Wilson, Zwicky believed that the smaller telescope's ability to cover wide swaths of the sky would uncover many new phenomena, such as dwarf galaxies, that had been missed with the deep but incomplete coverage of the larger telescope.

Zwicky's telescope was installed on Mount Palomar in 1936. It was the first telescope at Palomar Observatory, and the only operational telescope at the site between 1936 and 1948. Within months, Zwicky discovered two dwarf galaxies in the constellations of Leo and Sextans. In the course of a four-year survey, he found several more. Harlow Shapley of Harvard College Observatory used telescopes in South Africa to find more dwarf galaxies, and Walter Baade then used the hundred-inch telescope at Mount Wilson to confirm the existence of dwarf galaxies. He showed that they are less than a million light-years away and are most likely satellites of our Milky Way galaxy. Baade, in a paper coauthored with Hubble, referred to the objects as "dwarf nebulae," a reversal from Hubble's earlier position that dwarf galaxies do not exist. (Hubble was obviously a good enough scientist to recognize that they do indeed exist when presented with the evidence. However, there is no mention of Zwicky's work in the paper.) Today more than twenty dwarf-galaxy satellites of the Milky Way are known, and it is suspected that there may be more hidden in the shadow of the Milky Way's disk. Andromeda has a similar number. Not only do they exist, but they are also the most common galaxy type in the universe.

It is difficult to find and measure the masses of supermassive black holes in dwarf galaxies. The galaxies themselves are faint, and the signal of a concentration of mass in the central region of a faint galaxy is fainter still. The best way—indeed, the only feasible way at present—to locate them is to look for radiative signatures of a black hole that is accreting gas. By combining optical and X-ray observations, Amy Reines of the University of Michigan and her colleagues have found several. The most interesting one is in the galaxy RGG 118, located about 340 million light-years from Earth After their discovery in 2013 of evidence of a supermassive black hole in RGG 118 in data from the Sloan Digital Sky Survey, Reines and colleagues used Chandra and the 6.5-meter Clay Telescope in Chile for X-ray and optical follow-up observations. In a 2015 paper, they showed that the optical data provide evidence of gas swirling around an object in the center of the galaxy, and the X-ray observations pinpoint an X-ray source in the same area. The combined data indicate that a black hole with a mass fifty thousand times the mass of the Sun is located in the center of the galaxy, making it the smallest supermassive black hole ever detected in the center of a galaxy.

"It might sound contradictory, but finding such a small large black hole is very important," said Vivienne Baldassare of the University of Michigan, first author of a paper on the black hole in RGG 118. "We can use observations of the lightest supermassive black holes to better understand how black holes of different sizes grow" (2015). Indeed, the discovery of the RGG 118 black hole provides evidence that formation pathways must exist that produce black hole seeds of fifty thousand solar masses or less. Whether the RGG 118 black hole is typical or an anomaly is critical to understanding how supermassive black holes thousands of times more massive than this one form. Reines and others are hard at work to find more examples.

Meanwhile, other astrophysicists continue to search the distant reaches for those exceedingly faint primordial black hole seeds. One technique is to stack, or add up, X-ray counts found at the location of each galaxy in the field to treat the collection as a single target. Two groups, one led by Lennox Cowie of the University of Hawaii and the other by Ezekiel Treister of the Universidad de Concepción in Chile, have applied this technique to X-ray data in Chandra Deep Field South. They stacked the X-ray data at the positions of distant (observed as they were more than 12.7 billion years ago) quasars found in Hubble images. They found that the average supermassive black hole 12.7 billion years ago would have been radiating at one hundred times the luminosity of the RGG 118 black hole and thus has gone undetected. However, as the simulations of Pacucci and his colleagues show, the gas density around black holes during the first billion years after the Big Bang was probably considerably higher, so the black holes had more gas to accrete and should have been more luminous. The prospects seem good for detecting primordial black hole seeds in the not too distant future.

With rare exceptions involving merging galaxies with double nuclei, super-massive black holes are located in the nuclei of galaxies. This is not an accident. The nucleus is where the concentrations of stars, gas, and dust are greatest, so if supermassive black holes were going to form anywhere, it would be there.

Exactly how they got there and how they grew is currently one of the most active fields in astrophysics. Several plausible processes have been identified. Major mergers involving collisions between two approximately equal-sized galaxies can result in the channeling of gas into the supermassive black holes, which eventually merge. Supermassive black holes may also grow by more gradual processes such as the accretion of filaments or clumps of gas in and around the host galaxy, or through minor mergers involving satellite galaxies. Computer simulations indicate that in these cases the infall of gas or small galaxies into the main galaxy creates havoc in the main galaxy's gas, leading to the flow of matter into the central black hole. A growing body of evidence indicates that all of these processes play a role at different stages in the evolution of galaxies and their super-massive black holes.

Supermassive black hole mergers were common billions of years ago, with the peak activity occurring about 2 to 6 billion years after the Big Bang, or at roughly the same time that galaxies were forming stars at a furious pace. At this time, the universe was only about one-fourth to one-half of its present size, so collisions between galaxies would have been much more frequent. The finite speed of light means that when we look at extremely distant objects, we actually are looking back at extremely remote times and observing these events as they occur. Because of the vast distances, it is difficult to study these events in much detail, but mergers still occur in the current epoch, so we can examine these events to get a better idea of what they must have been like back then.

Chandra has observed various stages of the merger process. VV 340, also known as Arp 302, consists of two galaxies, seen in the early stages of their interaction. The galaxies are about thirty-three thousand light-years apart and will merge in a few million years, although it will likely take a few hundred million years for their supermassive black holes to merge into a single, larger black hole.

The galaxies that are collectively called SDSS J1254+0846 were discovered in the Sloan Digital Sky Survey (figure 39). Follow-up observations with Chandra and optical telescopes at Kitt Peak National Observatory in Arizona, Palomar Observatory in California, and the Magellan Telescope at Las Campanas Observatory in Chile by Paul Green of the Center for Astrophysics and his colleagues showed that SDSS J1254+0846 contains a binary quasar in the midst of a

39 | BINARY QUASAR

A Chandra image (blue) showing two quasars in galaxies SDSS J1254+0846, whose projected separation is about seventy thousand light-years, combined with an optical Magellan image (yellow) showing long tails around the host galaxies. These tails are produced by gravitational tides generated by the collision of two galaxies.

galaxy merger. The galaxies are slightly farther apart than the VV 340 ones, at seventy thousand light-years, but the supermassive black holes are more active and are shining brightly in X-rays. This discovery marked the first time a luminous pair of quasars has been clearly seen in an ongoing galaxy merger. The tails seen in the image are produced by enormous tides generated in the merger as the galaxies tug on each other. The mergers have likely triggered an increased flow into the central supermassive black holes and fueled the intense X-ray emission.

Chandra's image of NGC 6240 (figure 40), a butterfly-shaped galaxy that is the product of the collision of two smaller galaxies, shows two active supermassive black holes—the bright pointlike sources in the image—about three thousand light-years apart. In a few hundred million years, the black holes will merge to form one larger supermassive black hole. NGC 6240 is also a prime example of a massive galaxy in which stars are forming at an exceptionally rapid rate due to the merger.

The nearest known pair of supermassive black holes, at a distance of 160 million light-years from Earth, is in a spiral galaxy similar to the Milky Way. The black holes are located near the center of the spiral galaxy NGC 3393 (figure 38, page 120) and are heavily obscured. The obscuration makes them difficult to observe with an optical telescope, but the X-rays shine through, and Chandra observations reveal clear signatures of the pair of supermassive black holes. Separated by only 490 light-years, they are likely the remnant of a merger of two galaxies of unequal mass a billion or more years ago. Both black holes are actively growing, generating X-ray emission as gas falls toward them. The merger of two equal-sized spiral galaxies produces a galaxy with a distorted appearance that is the site of intense star formation, along with a pair of supermassive black holes. NGC 6240 is a good example of this process. However, NGC 3393 looks to be a well-organized spiral galaxy. Its central bulge is dominated by old stars, which is unusual for a galaxy containing a pair of black holes.

Computer simulations of the evolution of galaxies and supermassive black holes indicate that minor mergers involving a small galaxy and a larger one should be important to the formation of supermassive black holes. However, good candidates have been difficult to find because the merged galaxy is expected to look so typical. In fact, Giuseppina Fabbiano of Harvard-Smithsonian and her colleagues, who in 2011 reported the discovery of a black hole merger in NGC 3393, had no inkling that the double nucleus existed when they undertook a detailed investigation of the nucleus using special processing techniques. Their intent was to set better limits on the single supermassive black hole they thought was there. In NGC 3393, the two galaxies appear to have merged without a trace of the earlier collision, apart from the two black holes. This suggests that NGC 3393 may be the first known instance of an important pathway to the formation of a supermassive black hole. More examples are needed, but they are predicted to be rare in today's universe, since most mergers are a thing of the past.

"We may have been lucky," Fabbiano remarked in a NASA press release in 2011. "However, I believe it's healthy as an observer to keep an open mind and not be unduly influenced by theory. One thing we have learned from NGC 3393 is that normal-looking large bulge galaxies, not just clearly disturbed mergers, may be good hunting grounds for closely interacting merging active supermassive black holes. We intend to continue pursuing this hunt with Chandra."

Other than having a pair of supermassive black holes swirling around in its nucleus, NGC 3393 seems remarkably unremarkable. This is in contrast to NGC 6240, which is the site of vigorous star formation, likely triggered by the galactic merger. Perhaps NGC 3393 is relatively serene because it was the site of a minor merger that did not much affect the overall structure of the galaxy. Or perhaps the galaxy is not forming stars because there is not much gas left to form them from as the galaxy was cleared out by the energy output from the black holes. Computer simulations indicate that this sort of negative feedback can occur, and observations provide some striking evidence to support this idea.

40 | MERGING BLACK HOLES

NGC 6240, a galaxy 400 million light-years from Earth that holds two merging black holes. X-ray data from Chandra (red, orange, and yellow) is combined here with a Hubble optical image. The black holes, only three thousand light-years apart, are the bright points to the left of center in this image.

41 | CENTAURUS A

A Chandra image. The lowest-energy X-rays from this elliptical galaxy, 11 million light-years from Earth, are in red. Medium-energy X-rays are green, and the highest-energy ones are blue. An X-ray jet produced by a beam of high-energy particles from a supermassive black hole extends from the middle to the upper left for thirteen thousand light-years. A smaller counter-jet extending to the lower right can also be detected, as well as an arc of X-ray emission farther out. The Chandra image also highlights a dust lane that wraps around the galaxy's waist, a feature that is thought to be a remnant of a collision between Centaurus A and a smaller galaxy about 100 million years ago.

14

If a car was as fuel-efficient as these black holes, it could theoretically travel over a billion miles on a gallon of gas.

Christopher Reynolds, NASA press release, 2006

GREEN
BLACK HOLES

B lack holes have a reputation for being the ultimate sinks of matter and energy. To a certain extent, that reputation is well deserved. Matter falling beyond the event horizon of a black hole cannot escape, and it is forever out of touch with the rest of the universe. However, this is not the whole story.

For stellar mass black holes, the matter consumed by the black hole comes from a companion star. For a supermassive black hole, the menu is much more diverse. It can be gas that has escaped from stars in the neighborhood, which can encompass several hundred light-years. Or it can be comets or planets that stray too close to the black hole. Stars, too, can become food. The pull of the supermassive black hole's gravity is greater on the side of the star near the black hole than on its far side. These gravitational tidal forces can stretch the star into a long sausage shape and ultimately pull it apart. After the star is destroyed, the black hole's strong gravitational force pulls most of the remains of the star toward it. This infalling debris is heated to millions of degrees Celsius and generates a bright

X-ray flare, which gradually fades over a few years as the material falls beyond the black hole's event horizon.

A familiar—but fortunately much milder—example of tidal force is the differing gravitational force exerted by the Moon on the near and far sides of Earth. This force differential, or tidal force, stretches Earth out of round and produces lunar ocean tides. Several astrophysicists first discussed the possibility of tidal disruption of stars by supermassive black holes forty years ago. However, the first strong evidence for stellar death by black holes came in the form of a powerful X-ray outburst from the center of the galaxy RXJ1242-11. The outburst was tracked by Chandra, XMM-Newton, and the German Roentgensatellite, which is no longer operational. Other bright flares have been seen from galaxies, but this is the first one that occurred when Chandra and XMM-Newton were in orbit. Chandra located the event precisely, and XMM-Newton measured the spectral distribution of the X-rays emitted in the outburst. The results of these combined observations fit with a tidal disruption explanation, and they enabled researchers to rule out other possibilities for the X-ray outburst. One of the brightest flare-ups ever detected in a galaxy, it was deemed to have been caused by gas from the destroyed star that was heated to millions of degrees Celsius before being swallowed by the black hole.

The supermassive black hole in RXJ1242-11 is estimated to have a mass of about 100 million Suns. In contrast, the destroyed star likely had a mass about equal to the Sun. "This is the ultimate David versus Goliath battle, but here David lost," said Gunther Hasinger of the Max Planck Institute for Extraterrestrial Physics in Germany, one of the coauthors of the study. An estimated 1 percent of the star's mass was consumed by the black hole, and the rest was flung away in an X-ray-emitting wind.

In late 2014, a tidal disruption event was detected in the center of the galaxy called PGC 043234, near a supermassive black hole that is estimated to weigh a few million times the mass of the Sun. PGC 043234 is about 290 million light-years from Earth, so the event was the closest tidal disruption discovered in a decade. The event, named ASAS-SN-14li, was originally discovered in an optical search by the All-Sky Automated Survey for Supernovae (ASAS-SN) in November 2014. Chandra, Swift, and XMM-Newton were all able to get a look (figure 42).

The X-ray spectrum and the change of the X-ray emission over time indicate that the observed X-rays come from gas swirling in an accretion disk close to the event horizon of the black hole. As with many other accreting black holes, most of the matter does not fall into the black hole but is expelled outward in a wind. X-rays from the accretion disk are absorbed by the wind, imprinting an absorption "barcode" on the radiation from the accretion disk. Astrophysicists can use this barcode to examine the composition of the wind. It appears that the wind is composed predominantly of hydrogen and helium, with traces of oxygen, calcium, argon, sulfur, and other heavy elements.

What would happen if a star wandered too close to Sgr A*, the supermassive black hole in the center of our galaxy? The object G2 referred to earlier might have been such an object, but it stayed well away from the type of close encounter needed to disrupt a star. If a close encounter were to occur, the normally weak X-ray source associated with Sgr A* would flare up to be about fifty thousand times brighter than the brightest X-ray sources outside our solar system. At a distance of twenty-six thousand light-years, it would still be fainter than a strong solar flare, so it would not pose a threat to Earth.

In any event, the odds that a tidal disruption event will happen in a typical galaxy are low, about once every ten thousand years, so they are a rare meal for a supermassive black hole. That is not to say that we know what a typical meal for a supermassive black hole consists of, or how frequently they eat. Sgr A* is currently on a severe diet. Observations with the Submillimeter Array of radio telescopes atop Mauna Kea in Hawaii indicate that Sgr A* is consuming the equivalent of about half the mass of the Moon every year. The most likely source of the matter comes from winds produced by the dense collection of stars orbiting within a few to a few hundred light-years of the black hole. The size of the grazing area, or pasture, for the black hole depends on the black hole's gravitational force, which is proportional to the black hole's mass. Occasional clumps of matter, in the form of comets or asteroids, are likely on the menu. Near the black hole, it is possible for planets to get stripped away from their parent stars in close encounters with other stars by the strong gravitational tides of the black hole. Once in a great while, one of these stolen planets may fall into the supermassive black hole and produce a very bright flare.

The nearby (32 million light-years) galaxy NGC 3115 hosts a supermassive black hole with a mass estimated from the velocity of stars in the central region of the galaxy to be a few hundred times more massive than Sgr A*. This object provides the next best example to Sgr A* for studying the flow of matter into a quiescent supermassive black hole. A cloud of multimillion-degree Celsius gas is detected within a few hundred light-years of the black hole, but the X-ray source associated with the black hole is so feeble as to be undetectable, implying

42 | TIDAL DISRUPTION

A schematic of tidal disruption of a star by a black hole, with spectrum. The artist's illustration depicts material from a shredded star (the reddish-orange streak) being pulled toward the black hole in the source called ASAS-SN-14li. The X-ray spectrum in the inset provides information about the temperature, composition, and motion of gas swirling around the black hole. The spectrum also contains evidence that some X-rays from the accretion disk are absorbed by a wind flowing away from the black hole.

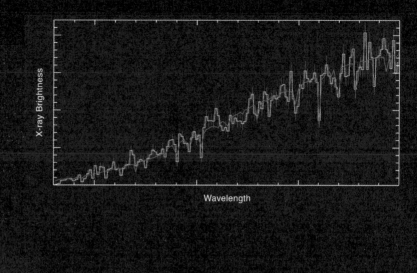

that very little of the infalling matter reaches the black hole. Rather, it must be expelled in a wind.

A pattern is emerging. Supermassive black holes exist in the centers of most and quite possibly all galaxies. If properly fed—make that fed to excess by some major event such as a large-scale merger with another galaxy—they can be prodigious sources of X-rays and other wavelengths of radiation. As such, they explain quasars and the radiation from the nuclei of Seyfert and other types of active galaxies. When the contributions of all these types of X-ray-emitting supermassive black holes are added up, they explain the diffuse X-ray glow that was apparent from the very first X-ray observations with counters aboard sounding rockets. But that is not the normal state of things. Most supermassive black holes, if fed normally, do next to nothing. They just sit there in the middle of the galaxy, quietly spinning around.

The spinning-around part is important, and I will return to it in a moment. For now, back to the question of the black hole's reputation as a sink of matter and energy. Chandra observations such as those described earlier reveal the rate at which gas is falling toward the supermassive black hole located in the center of a galaxy. These observations can be combined with estimates of energy flowing out of a galactic nucleus in the form of light, particles in high-velocity winds, and tornado-like jets, to get an idea of the efficiency of a black hole engine.

"Just as with cars, it's critical to know the fuel efficiency of black holes," Steve Allen remarked in a 2006 NASA press release; Allen has used Chandra to study black holes as well as galaxy clusters. "Without this information, we cannot figure out what is going on under the hood, so to speak, or what the engine can do."

Allen and his colleagues, including Chris Reynolds, used Chandra to pursue the question of the efficiency of black hole engines. They observed nine supermassive black holes at the centers of elliptical galaxies (figure 41, page 130). These black holes were carefully chosen so that the mass inflow and energy outflow could be accurately measured. The Chandra results showed that these "quiet" black holes are all producing much more energy in jets of high-energy particles than in visible light or X-rays. These jets create huge bubbles, or cavities, in the hot gas in the galaxies. The size of these cavities gives an indication of the power contained in the jets that created them.

The efficiency of the black hole "machine" was found to be about 2 percent of the total energy, including the $E = mc^2$ energy associated with the mass of the particles, flowing toward the supermassive black hole, with an indication that it might go even higher for more powerful black holes. This may not seem like much until you consider that it is three times greater than the efficiency of nuclear fusion of hydrogen to helium inside a star, twenty times more efficient than nuclear fission, and roughly 50 million times more efficient than burning coal. As Peter Edmonds of Harvard-Smithsonian has noted, "Super-advanced aliens would generate much more energy from sending a star into a black hole than letting it shine for billions of years!" (2005).

The process for extracting energy efficiently from stellar mass black holes was discussed in an earlier chapter. It involves a spinning black hole and a magnetic field, and it works the same way for supermassive black holes. As with the stellar black holes, magnetic fields and spin can make all the difference, enabling black holes to affect the entire galaxy they inhabit—and in some cases beyond. The combination of spin and magnetic fields creates enormous electric fields that fling matter away from the black hole at the last moment, making for some of the most powerful and efficient mass redistribution engines in the universe.

In the galaxy M87 (figure 43), Chandra observations imply that most of the matter never makes it to the black hole, so the mass accretion rate is only about one Jupiter per year. Yet this rate is apparently enough to fuel a powerful jet of high-energy particles, if the black hole is spinning rapidly. But is it? In M87, the evidence is indirect, but the spin rate has been measured for a few very distant supermassive black holes, thanks to gravitational lensing. Gravitational lensing has proven to be critical for studying the distribution of dark matter in galaxy clusters. It also provides a natural telescope, which is sometimes referred to as Einstein's telescope, for studying the accretion disks around distant supermassive black holes.

The quasar RX J1131-1231 (RX J1131 for short) hosts a supermassive black hole with an estimated mass of 200 million Suns. The X-rays from RX J1131 have traveled 6 billion years to reach Earth. As luck would have it, a massive foreground galaxy lies along the path of the X-rays. The bending of space by the gravity of stars in these foreground galaxies acts like a lens and magnifies the light from the quasar behind it (figure 44). The alignment is such that multiple images are produced, and the details of the radiation close to the black hole can be studied. As with the stellar black holes, the observed distortion of the X-radiation by gravity can be used to determine the size of the accretion disk and how fast it is spinning. Chandra and XMM-Newton data showed that the black hole is spinning at

87 percent of the maximum value. A similar result has been found for two other gravitationally lensed quasars.

The evidence shows that at least some black holes are spinning rapidly. This rapid spin causes almost all of the matter and energy that approaches the event horizon to be recycled back into the galaxy, and sometimes beyond, as winds, jets, or radiation. It turns out that black holes are, paradoxically, the ultimate fuel-efficient green machine.

43 | M87 GALAXY

A Chandra image of this galaxy showing a bright X-ray jet in its nucleus. The jet is produced by high-energy particles generated by magnetized gas spiraling in a disk toward a supermassive black hole. The jet is thought to be pointed at a slight angle to the line of sight, out of the plane of the image, so its actual length may be much greater than the projected length of five thousand light-years. On a larger scale, rings of hot gas are detected, likely remnants of repetitive explosive activity generated as gas cools and falls toward the supermassive black hole over the course of millions of years. The infall of gas may trigger an outburst that expels much of the gas, which will eventually cool, starting a new cycle.

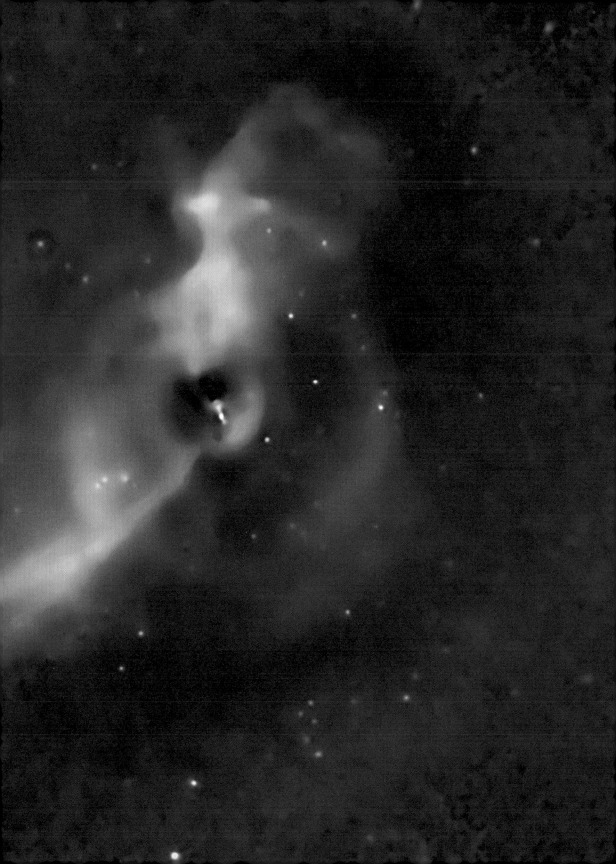

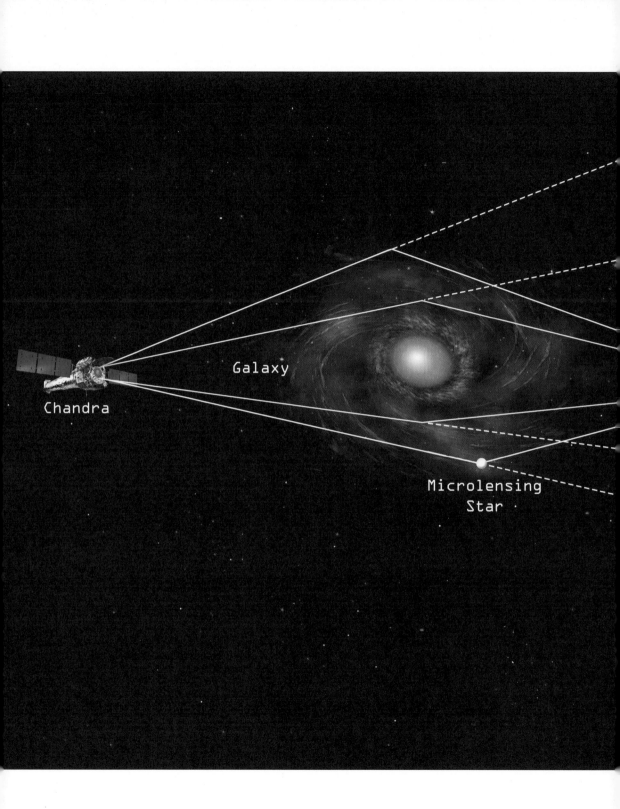

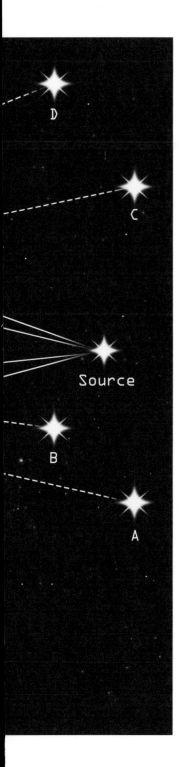

44 | GRAVITATIONAL MICROLENSING

A schematic. The gravitational field of stars in a foreground galaxy can bend and magnify light from a much more distant object to produce multiple images of the object, as shown here. In the case of the Cloverleaf quasar (labeled "Source"), Chandra has found four separate images (A–D) of this single object, which is 11 billion light-years away. These "lenses" make it possible to study the structure of accretion disks at a level of detail otherwise unattainable with even the most powerful telescopes.

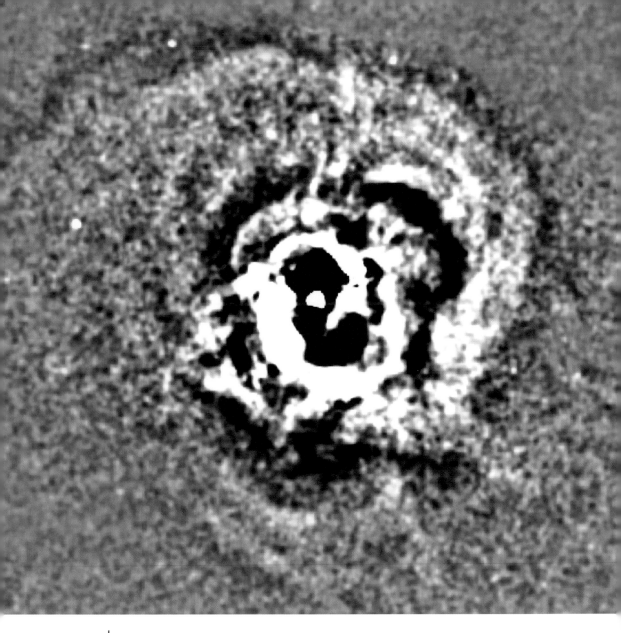

45 | PERSEUS GALAXY CLUSTER

A galaxy cluster about 250 million light-years from Earth. A special image-processing technique brings out changes in brightness in this Chandra image to reveal ripples in the hot gas that fills the cluster. These ripples are likely due to sound waves propagating through the hot gas following explosive activity connected to the supermassive black hole (the bright spot) at the center of the cluster.

15

It's important to have a feedback loop.

Elon Musk, 2014

BLACK HOLE
FEEDBACK

The jets of high-energy particles and associated electromagnetic funnels produced by spinning supermassive black holes are immensely powerful, affect-
ing the evolution of galaxies and even clusters of galaxies that span millions of
light-years. All this energy is from a powerhouse roughly the size of our solar system. Proportionally, it is as if an energy source the size of a blueberry could heat
a gas cloud the size of Earth to millions of degrees Celsius.

It seems mind-boggling, but many astrophysicists now concur that the
power from rotating supermassive black holes can be used to solve a long-
standing puzzle called the cooling flow problem. Recall that the hot gas in
galaxy clusters contains more mass than all the stars in all the hundreds, and
in some cases thousands, of galaxies in a cluster.

The gas, which is heated primarily by the slow gravitational collapse of
the cluster, has a temperature of tens of millions of degrees, and glows brightly
in X-rays. Optical telescopes cannot see it, and X-rays cannot penetrate Earth's

atmosphere, so the discovery and study of this gas has depended on orbiting observatories. Two decades ago, astronomers observing with NASA's Einstein X-Ray Observatory and other instruments noticed that the X-rays carry away so much energy that the gas should cool off and settle into the center of the cluster— thus the term "cooling flow." Andrew Fabian and his colleagues led the way in investigating these flows with Einstein and the follow-up ROSAT X-ray satellite. They showed that if the flows persisted for a billion years, the gas deposited in the central regions of the cluster could form trillions of new stars.

The only trouble was, no one could find these stars. Observers looked in vain for large amounts of cool gas and hordes of newly formed stars. If a black hole had swallowed them all, it would weigh as much as a trillion stars, and not even the biggest black hole is that massive. Some astrophysicists maintained that in this case, the absence of evidence was evidence of absence, and that large-scale, long-term cooling flows do not exist. A possible explanation was that the central galaxy of the cluster somehow produces energetic outflows that heat the gas enough to offset the radiative cooling. Radio astronomers had for years been accumulating evidence for such outflows, yet it was uncertain whether the out-flows could provide enough energy distributed over a large enough volume to halt the cooling flows.

Resolving the cooling flow problem was one of the goals of both Chandra and XMM-Newton. Because the gas radiates away its energy fairly slowly, it preserves a record of the past activity in the clusters over the last few billion years. For instance, it retains the elements and energy injected into it by supernova explosions in the cluster galaxies. Like archaeologists unearthing the past, astronomers have used the new telescopes to excavate the relics present in galaxy clusters and piece together their histories.

The brightest cluster in X-rays is the Perseus cluster (figure 45, page 142). It has a high intrinsic luminosity, and, at a distance of about 250 million light-years, it is relatively nearby. In the 1990s, ROSAT discovered two vast holes in the X-ray emission in the central fifty thousand light-years of the cluster, around the giant galaxy NGC 1275. With Chandra, Fabian and his colleagues went back for a closer look. The Chandra image showed the cavities in exquisite detail. When the radio data were added to the Chandra image, the jets fit nicely inside the cavities.

The X-ray cavities are almost empty of hot gas, but they are not altogether empty. They are filled with high-energy protons and electrons and magnetic fields. These bubbles have a lower density than the hot gas, so they rise buoyantly, like bubbles in boiling water, and push aside the hot gas.

By special processing of the Perseus cluster images, as seen in figure 45, Fabian's team unveiled a series of nearly concentric ripples. The density and pressure of the gas, though not its temperature, change abruptly at the innermost ripple, indicating that this feature is a weak shock wave. At the outer ripples, the density and pressure vary gradually, indicating that these ripples are sound waves. The spacing of the ripples (thirty-five thousand light-years) and the calculated speed of sound in the gas under these conditions (2.6 million miles per hour) imply that 10 million years passed between the events producing the ripples. The pitch of the sound waves translates to a note of B-flat, fifty-seven octaves below middle C on a piano.

In the Virgo galaxy cluster, the nearest cluster to us at a distance of about 50 million light-years, William Forman of Harvard-Smithsonian and his colleagues observed the central dominant galaxy in this cluster, M87, using Chandra. In addition to the jet from the black hole, they found a web of filamentary structures farther out, each about a thousand light-years across and fifty thousand light-years long. The filaments may be formed from a series of buoyant bubbles arising from outbursts spaced about 6 million years apart. Forman's team also detected a ring of hotter emission with a radius of about forty thousand light-years, probably a weak shock front, as well as a large X-ray cavity about seventy thousand light-years from the galaxy's center. The sound waves in this case might be about an octave higher in pitch than those of the Perseus black hole.

Similar features show up in other clusters, and in groups of galaxies as well. Groups are smaller versions of their cosmic cousins, galaxy clusters. Instead of containing thousands of galaxies, as clusters do, they typically have a handful of large galaxies and a few dozen dwarf galaxies. Many, but not all, groups also contain huge clouds of multimillion-degree gas.

The galaxy NGC 5813 is part of a group about 100 million light-years from Earth (figures 46–47). The group contains another large galaxy, NGC 5846, and more than two hundred smaller galaxies clustered around the two larger galaxies. Chandra observations of the region around the NGC 5813 subgroup reveal evidence that multiple eruptions have occurred in this group over the course of the past 50 million years. Three pairs of colinear cavities with bright rims of X-ray emission were detected at three thousand, twenty-five thousand, and sixty-four thousand light-years from the central supermassive black hole, which has a mass of 280 million solar masses. The cavities all had bright rims characteristic of shock waves moving through the hot gas at about Mach 1.5, or 1.5 times the speed of sound. The shock speeds, together with the size of the cavities, imply that distinct outbursts occurred about 2 million, 15 million, and 50 million years ago. The shock energy appears to be sufficient to balance the radiative cooling of the hot gas.

Chandra observations by Brian McNamara of the University of Waterloo in Canada and his colleagues revealed X-ray cavities with associated radio emission in the Hydra A, Hercules A, and Abell 2597 clusters. They also discovered bubbles that were faint in both radio waves and X-rays, indicating that the energetic particles in them have dissipated most of their energy. These "ghost cavities" are displaced well away from the central black hole and may be relics of past activity.

McNamara and colleagues also used Chandra to discover the most spectacular cavities yet. They are in the cluster MS 0735.6+7421 (figure 48). Although the image is not as detailed as the one of Perseus, it tells an amazing story. Each of the two opposing X-ray cavities is about 640,000 light-years across—nearly seven times larger than the disk of our Milky Way galaxy. The size of the cavities, and the observed density and temperature of the gas around them, indicates they are 100 million years old and contain as much kinetic energy as 10 billion supernovas.

John Peterson of Purdue University and his colleagues used XMM-Newton to make a comprehensive study of clusters with and without cooling flows. Their research shows that cooling flows do not occur in clusters that have such high-energy bubbles. Somehow the bubbles are transferring heat to the gas and preventing it from cooling. Exactly how they are doing this is still an open question. Various proposals—shock waves, heat conduction from the hot outer gas, heating by the high-energy particles in the bubbles, or the decay of turbulent eddies—have been floated, but no one mechanism has gained acceptance.

But the biggest question of all—namely, where the energy came from to begin with—has been answered. A rapidly spinning supermassive black hole can do it. Calculations show that black hole jets have two major components: a matter-dominated outflow that moves at about a third of the speed of light, forming the outer sheath of the funnel, and an inner region along the axis of the funnel that contains a rarified gas of extremely high-energy particles. It is the inner region that carries much of the energy and creates the dramatic structures observed in Perseus and other galaxy clusters.

One of the most astounding features of jets is the pencil-thin shape that they can maintain despite traveling hundreds of thousands of light-years, far beyond the confines of their parent galaxies. The pressure of the gas near the black hole can get a jet started as a narrow beam, and it may be that inertia keeps the jet narrow, much like a blast of water from a hose or steam from a high-pressure teakettle. The tightly coiled magnetic field that is spun out with the jet may also play a role.

46 | NGC 5813 GALAXY

An optical image of the dominant galaxy in a group of a few dozen galaxies about 100 million light-years from Earth.

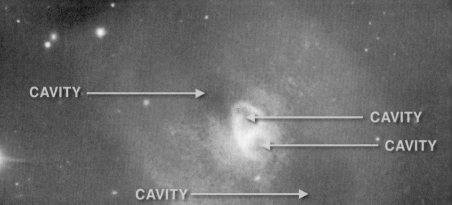

Regardless of the confining mechanism, the pressure of the rarified gas through which jets move gradually takes its toll. The jets slow down and billow out, creating enormous magnetized clouds of high-energy particles. These clouds continue to expand, pushing out the surrounding gas to create the dark X-ray cavities observed by Chandra (figure 49).

This sequence of events—gas falls into a rapidly spinning black hole to form megajets that carve out gigantic bubbles of high-energy particles and heat vast volumes of space—is feedback of truly cosmic proportions. The black hole is both responding to and influencing events on the scale of the entire galaxy cluster.

A likely scenario is as follows. Initially the gas in the cluster is very hot, and the supermassive black hole in a centrally located large galaxy is quiet. Over about 100 million years, gas in the central region of the cluster cools and drifts toward the central galaxy in a cooling flow where it can form stars.

Some of the gas in this cooling flow condenses into stars that become part of the central galaxy, and some sinks farther inward to feed the supermassive black hole. In so doing, it creates an accretion disk and activates high-power jets. The jets blast through the galaxy and out into the cluster gas, where they heat the gas. The cooling flow is greatly diminished, if not shut off altogether. The supply of gas to the supermassive black hole declines. Without an abundant gas supply, the supermassive black hole gradually becomes dormant and the jets fade away, leaving the cluster gas without a heat source. Millions of years later, the hot gas in the central region of the cluster cools sufficiently to initiate

47 | BUBBLES IN NGC 5813 GALAXY GROUP

A composite image of the NGC 5813 group of galaxies showing X-rays detected by Chandra (purple) from hot gas in the group, along with an optical image (red, green, and blue) from the Sloan Digital Sky Survey. The cavities, or bubbles, carved out of the X-ray-emitting hot gas that envelops the galaxies indicate that repeated eruptions from a supermassive black hole have occurred over a period of 50 million years.

a new season of growth for the galaxy and its supermassive black hole, and thus the cycle continues.

This scenario is supported by high-resolution X-ray and radio images of the NGC 5813 group, as well as Virgo, Perseus, Hydra, and other clusters, which show evidence of repetitive outbursts from the vicinity of the central galaxies' supermassive black holes. Magnetized rings, bubbles, plumes, and jets ranging in size from a few thousand to a few hundred thousand light-years strongly suggest that intermittent violent activity has been ongoing in these clusters for hundreds of millions of years.

One of the most spectacular examples of the star formation part of the process is the brightest galaxy in the Phoenix Cluster, which is seen at a lookback time of 5.7 billion years (figure 50). This system has the mass of about 500 billion Suns, and it produces more X-rays than any other known cluster. Over the past few million years, stars have been forming in this galaxy at the astounding rate of six hundred solar masses per year, compared to a rate of perhaps one per year for the Milky Way galaxy. Chandra data also reveal the presence of deep X-ray cavities in the inner thirty thousand light-years of the 50-million-degree Celsius gas cloud that envelops the central region of the cluster, an indication that the cycle of infall, blowback, cooling, and so forth has been ongoing for some time.

48 | MS 0735.6+7421 GALAXY CLUSTER

The site of one of the most powerful black hole eruptions ever observed. X-rays detected by Chandra (blue) show the 50 million-degree Celsius gas that composes most of the cluster's baryonic mass. The Chandra data also reveal cavities in the hot gas. Each one is roughly 640,000 light-years in diameter, or nearly seven times the diameter of the Milky Way, and they were created by an outburst from a supermassive black hole at the cluster's center. The outburst ejected huge jets (pink), which were detected by the Jansky Very Large Array radio telescope. These data, combined with Hubble optical data, show the cluster's galaxies and foreground stars in the field of view (orange). The distance to MS 0735.6+7421 is about 2.6 billion light-years.

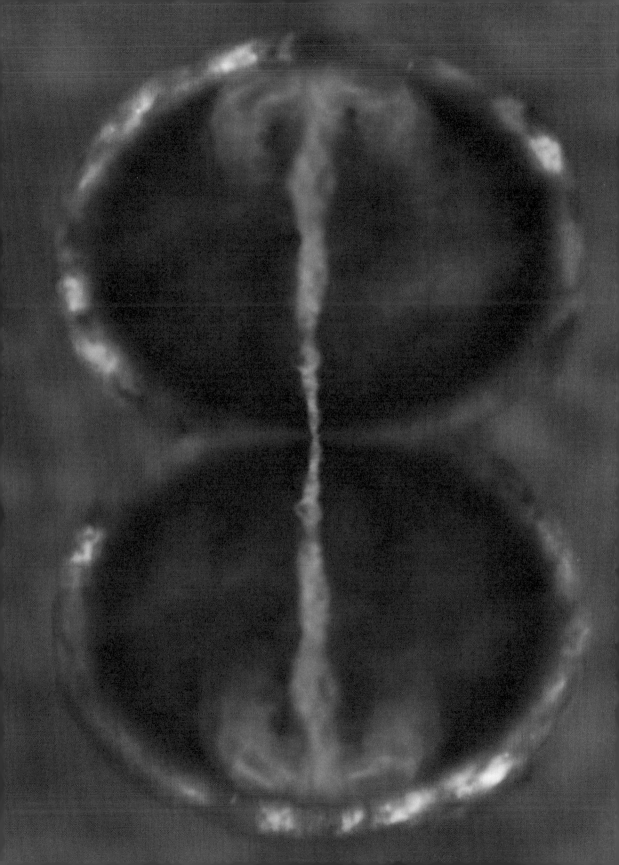

One startling implication is that the supermassive black holes in the centers of these clusters are still growing at a rapid rate even to the present day. Astrophysicists had thought that the growth of supermassive black holes had tapered off a few billion years ago. In the case of cluster MS 0735, the activity indicates that the supermassive black hole has gulped down the equivalent of 300 million suns in the past 100 million years—nearly doubling in size and mass over that relatively brief interval. Yet the central black hole shows no other signs of activity, such as bright X-rays or visible light, which are usually emitted by active holes. It is only through the X-ray cavities that we can discern the properties of this amazing system.

Galaxy collisions, an ever-present hazard in the central regions of galaxy clusters, can rekindle a supermassive black hole's activity by giving it another generous serving of matter. A smaller galaxy passing too close to the giant central galaxy is torn asunder. Its stars are assimilated, some of its gas goes down the black hole drain, and its own central black hole merges with the one in the giant galaxy. The enormous cavities observed in MS 0735 were probably the end result of a sequence of events initiated when a merging galaxy caused a huge influx of gas into a supermassive black hole.

The role of collisions may help scientists to understand the evolution of galaxies in the early universe, when galaxies were closer together and mergers of galaxies were common. A growing body of research indicates that many aspects of galaxy formation and evolution—the size of galaxies, the shapes of galaxies, the rate of star formation—can be understood in terms of a cosmic cycle involving mergers of galaxies. Large-scale computer simulations by Philip Hopkins of Caltech and his colleagues show that mergers of gas-rich galaxies trigger bursts of star formation and inflow of gas into the central region. The inflowing gas fuels rapid growth of the supermassive black hole and intense radiation from its vicinity. Blowback ejects much of the gas from the galaxy, star formation rapidly slows, and accretion onto the black hole declines—until another merger occurs.

Most of the black hole feedback that shaped the evolution of galaxies occurred about 8 to 10 billion years ago, when they were young. This epoch can be observed only in distant galaxies, which makes the study of details such as

49 | HIGH-ENERGY JETS

The action of magnetic fields, combined with gas spiraling around a supermassive black hole, can produce energetic jets of gas flowing away from the black hole. These jets push gas away to create the dark cavities observed in the cluster's gas cloud.

jets, bubbles, and waves difficult. Fortunately, similar (though not identical) blowback processes are occurring in relatively nearby galaxy clusters, where Chandra can directly image and measure them.

It may seem strange that supermassive black holes, objects with masses that range from a few million to hundreds of millions of solar masses, can have such a large impact on galaxies whose masses range from a few billion to a few hundred billion solar masses, let alone galaxy clusters with masses measured in the hundreds of trillions of solar masses. The reason is the concentrated nature of supermassive black holes and their gravitational fields. Supermassive black holes are by far the largest supply of gravitational potential energy in an entire galaxy. By tapping this energy through accretion disks and the launching of powerful megajets, black hole feedback is one of the most important processes at work in the universe.

So maybe black holes aren't all bad. In fact, maybe "bad is the new good," as the song goes.

50 | PHOENIX CLUSTER SPT-CLJ2344-4243

An X-ray and optical composite. The X-ray image (blue) is from Chandra, and the optical image (red and yellow) is from Hubble. The bright central galaxy appears in white in both the X-ray and the optical images. Chandra data reveal large cavities in the X-rays, which are thought to have been carved out of the surrounding gas by powerful jets of high-energy particles emanating from the region near a supermassive black hole in the cluster's central galaxy. Massive filaments of gas and dust, extending 160,000 to 330,000 light-years, surround the X-ray cavities. The image is about 1.5 million light-years across, and the distance to the Phoenix cluster is 5.7 billion light-years.

PART III

THE BEA

UTIFUL

51 | CAS A SUPERNOVA REMNANT

A Chandra image showing low- (red), medium- (green), and
high-energy (blue) X-rays. The forward shock wave is seen in blue, the
stellar ejecta heated by the reverse shock in red and green. The neutron
star discovered in the first Chandra image is the little white seed-shaped
spot near the center. The image is about twenty-nine light-years on
each side.

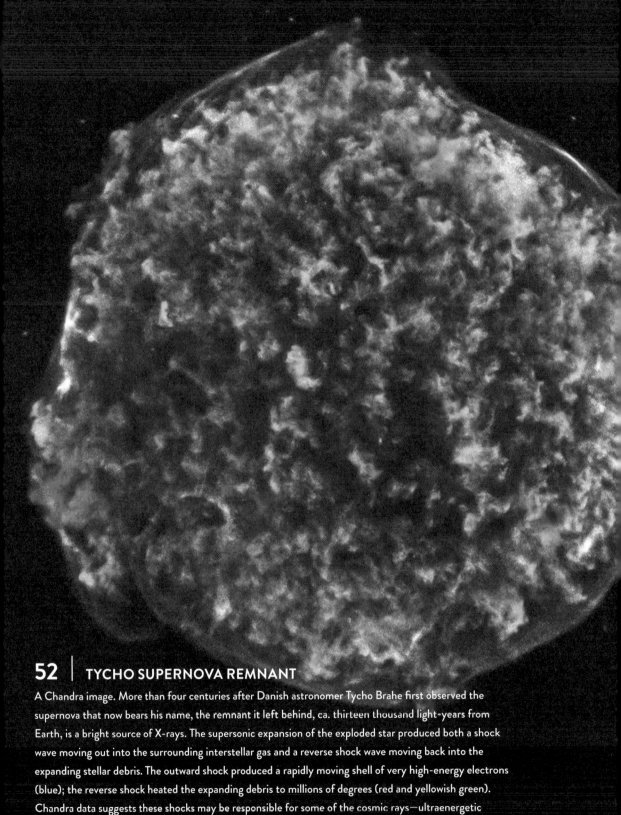

52 | TYCHO SUPERNOVA REMNANT

A Chandra image. More than four centuries after Danish astronomer Tycho Brahe first observed the supernova that now bears his name, the remnant it left behind, ca. thirteen thousand light-years from Earth, is a bright source of X-rays. The supersonic expansion of the exploded star produced both a shock wave moving out into the surrounding interstellar gas and a reverse shock wave moving back into the expanding stellar debris. The outward shock produced a rapidly moving shell of very high-energy electrons (blue); the reverse shock heated the expanding debris to millions of degrees (red and yellowish green). Chandra data suggests these shocks may be responsible for some of the cosmic rays—ultraenergetic particles—that pervade the galaxy and bombard Earth. (See also figure 73.)

16

Rage, rage against the dying of the light.

Dylan Thomas, 1952

GOING NOT GENTLE INTO THAT GOOD NIGHT

In centuries past, there was a belief that the stars governed our fate through "cosmic sympathy." Occurrences in the Macrocosm, or the heavens, were thought to presage fateful events in the Microcosm, such as success or failure in battle or the death of a powerful ruler. For this reason, court astrologers took their work seriously, and well they should have. Their lives literally depended on it: They could be executed for failure to predict an eclipse. Today's astronomers are beneficiaries of the meticulous recordkeeping of ancient Chinese, Japanese, Korean, Arabic, and European observers.

An excerpt from the chronicles of the Sung dynasty contains an account of the observation of the supernova of 1054 CE that produced the Crab Nebula. This account hints at the flavor of the times and the conditions under which the court astrologers worked: "Prostrating myself, I have observed the appearance of a guest star; on the star there was a slightly iridescent yellow color. Respectfully, according to the dispositions of the Emperor, I have prognosticated, and the result said: The guest star does not infringe on Aldebaran; this shows that the Plentiful One is Lord and that the country has a Great Worth" (Duyvendak 1942). If you take into account the passage of ten centuries and the developments in science and technology since then, this report could pass for a progress report on a research grant from the National Science Foundation or NASA.

Interdisciplinary research on the reports of astrologers by historians and astronomers provides a valuable record of the most violent events that have occurred in our galaxy over the past two millennia. Current estimates suggest that about three dozen supernovas should have occurred in the galaxy over that time. The relative scarcity of reported supernovas may be due to several factors. Many of the events could have been in the southern sky, others could have occurred on the far side of the galaxy, and some might have been embedded in obscuring clouds of dust and gas.

The records contain many reports of "new stars," "guest stars," and "stellae novae," but only a few of them were supernovas. The supernovas are distinguished by two criteria: (1) the duration of time over which they could be observed (at least three months, and as long as three years in the case of the supernova of 1006 CE; see figure 72, pages 206–07), and (2) by the discovery in modern times of expanding supernova remnants of hot gas and energetic particles at the locations of the reported supernovas.

The brightest supernova observed was the supernova of 1006 CE, which was brighter than Venus and visible for several years. However, the remnant of SN 1006 has not attracted nearly as much attention as the supernova of 1054 CE. This is because SN 1006 was likely produced by a thermonuclear explosion that completely disintegrated a white dwarf star (a Type Ia explosion) and left nothing behind except the expanding stellar debris. In contrast, the Crab Nebula was the product of the collapse of the core of a massive star. The collapse triggered a supernova that ejected most of the star, but left behind a rapidly spinning neutron star. This cosmic dynamo produces a tornado of magnetic fields and high-energy particles that continue to light up the expanding debris. The detection by Chandra of rapidly spinning neutron stars in supernova remnants is proving to be one of the best ways to distinguish between the types of supernova explosions, and thereby to understand the nature of the star that exploded.

All the historical supernovas occurred thousands of light-years away, so their physical effects on Earth were negligible. However, one supernova played an important role in the development of modern physics and astronomy.

The supernova of 1572 (see figure 52, page 158) occurred before the invention of the telescope, before Galileo and Newton, at a time when the concept of an exploding star was less plausible than witchcraft. An exploding star was contrary to the order of nature, according to which stars were symbols of the eternal and unchangeable, part of a realm of permanence located above the ever-changing, ever-corruptible world below.

Then, in November 1572, a star brighter than the planet Venus appeared suddenly in the constellation of Cassiopeia. Make no mistake, these people did not know calculus, because it hadn't been invented yet, and they had some strange ideas about what happened on Halloween, but they knew the sky far better than most of us. The new star, or *nova*, to use the Latin, of 1572, was noticed throughout Europe and in Asia. The appearance of the nova of 1572 and the subsequent detailed description of its properties by Tycho Brahe played a major role in revolutionizing astronomy, and all of science.

Tycho, as he was known, was a stormy, boisterous astronomer known for his acid tongue and brass nose. Where he got the acid tongue is a matter of speculation. What is known about his childhood suggest it could have been a result of either heredity or environment, or both. The brass nose (it is sometimes reported that Tycho's nose was silver, but in 2010 his body was exhumed to confirm that it was not) is another matter. It was acquired when Tycho lost his natural nose in a duel at age twenty, reportedly precipitated by an argument with a fellow student over who was the best mathematician.

Tycho, as it turns out, was neither a particularly good mathematician nor a skilled swordsman. But he was a great astronomer, probably the greatest of the age before the telescope. He built elaborate instruments for locating stars on the celestial sphere, kept meticulous records, and worked diligently. He noticed the nova of 1572 right away, and he immediately set to work making accurate measurements of its position in the sky relative to other stars in the constellation of Cassiopeia, where the star appeared. He showed that, although the brightness of the star declined steadily from being as bright as Venus until it became invisible, its position remained fixed. This proved that the new star was not a comet. It belonged to the "Eighth Sphere" of the fixed stars. The explosion of the star had shattered forever the widely accepted doctrine of the incorruptibility of the stars, and it set the stage for the work of Kepler, Galileo, Newton, and other eminent scientists of the time.

Fast-forward to 1934, when Fritz Zwicky and Walter Baade presented evidence for a previously unrecognized celestial phenomenon, which they called a supernova. Other stellar outbursts called novae were well known by then, and their properties fairly well described. In fact, Hubble had used the characteristics of novas to show that the Andromeda and other galaxies are millions of light-years from Earth and are vast assemblages of stars like the Milky Way galaxy.

Zwicky and Baade argued that supernovas are millions of times more energetic than ordinary novas, and each clearly represents an explosive event that involves an entire star, not just its outer layers. They and their colleagues undertook comprehensive surveys to search for supernovas. By the time the decade was out, their existence as a new type of phenomenon was well established, and Tycho's nova was upgraded as Tycho's supernova a few years later. In 1941, Rudolph Minkowski showed that supernovas could be divided into two types based on their optical properties. Type I explosions do not show evidence for hydrogen in the expanding debris ejected in the explosion, but Type II supernovas do show such evidence.

As other astronomers joined in the hunt for supernovas and the number of discovered supernovas increased exponentially, it became evident that Type II supernovas occurred in regions with lots of bright, young stars, such as the spiral arms of galaxies. They apparently do not occur in elliptical galaxies, which are dominated by old, low-mass stars. Since bright young stars are typically stars with masses greater than about ten times the mass of the Sun, this and other evidence led to the conclusion that Type II supernovas are produced by massive stars.

Type I supernovas were soon divided into subclasses Ia, Ib, and Ic. Types Ib and Ic turned out to be more like Type II supernovas, but Type Ia supernovas were different. Unlike the other types of supernovas, Type Ia supernovas generally occur in all types of galaxies, including ellipticals, and they show no preference for regions of current stellar formation. This means that they likely come from an older stellar population. Finally, the variation of the light from Type Ia supernovas with time, called the light curve, was very similar for all Type Ia events, and markedly different from the other types.

In 1945, Baade, in a remarkable bit of forensic astronomy, used Tycho's description of the supernova of 1572 to show that it fits nicely into the Type Ia category. A hunt for the supernova remnant proved fruitless until radio astronomers at Jodrell Bank in England found a strong source of radio waves near the position of Tycho's supernova. In the 1960s, the remnant of Tycho's supernova was finally detected as a faint nebula. In 1970, Paul Gorenstein of the Harvard-Smithsonian Center for Astrophysics and his colleagues discovered X-ray emission from Tycho's supernova remnant by using an X-ray detector aboard a sounding rocket. In 2008, Oliver Krause of the Max Planck Institute analyzed light echoes produced from the scattering of light off interstellar clouds near Tycho to view the supernova 436 years after the event and confirm that Tycho's supernova was indeed a Type Ia supernova.

In 1973, John Whelan and Icko Iben Jr. of the University of Illinois proposed that a Type Ia supernova is produced by the explosion of a white dwarf star. As the condensed remnant of what used to be a Sunlike star, a white dwarf star is a dense ball primarily composed of carbon and oxygen atoms. It is intrinsically the most stable of stars, as long as its mass remains below the so-called Chandrasekhar limit of 1.4 solar masses.

White dwarfs are among the dimmest stars in the universe. Even so, they have commanded the attention of astronomers ever since the first white dwarf was discovered in the middle of the nineteenth century. One reason for this interest is that white dwarfs represent an intriguing state of matter; another reason is that most stars, including our Sun, will become white dwarfs when they reach their final, burned-out, collapsed state.

For time out of mind, Sirius, the brightest star in the sky, has played a prominent and sometimes peculiar role in mythology and astronomy. In ancient Egypt, Sirius's hieroglyph, the dog, appeared on monuments and temple walls dating back beyond 3000 BCE. The rising of Sirius just before the Sun on the summer solstice signaled the beginning of summer, and coincidentally, the coming flooding of the Nile, an event critical for supplying the water necessary for agriculture. The temple of Isis-Hathor at Denderah and numerous other temples in the Nile Valley were oriented to the rising of Sirius. In ancient Greece and Rome, Sirius was thought to be a bad influence, responsible for drought, disease, and madness in dogs. We still speak of the dog days of summer in reference to the ancient association of Sirius with the withering heat of July and August.

The brightness of Sirius led the celebrated philosopher and cosmologist Immanuel Kant to speculate in the eighteenth century that Sirius is the center of our Milky Way galaxy, an intriguing but wrong idea. Sirius appears so bright because it is very near to Earth, at a distance of 8.7 light-years, and because it is intrinsically more luminous than an average star. Sirius was in the news—at least among astronomers—in the 1830s and 1840s when the astronomer and mathematician Frederick Bessel concluded that the apparent motion of Sirius was not a smooth arc, but an irregular curved path caused by an invisible companion. The theoretical orbit of this companion was computed, but a search for the companion to Sirius was fruitless.

Then in 1862, Alvan Clark, a renowned telescope maker, wanted to test out his latest creation, an 18.5-inch refracting telescope at the Dearborn Observatory of

Northwestern University. He decided to check it out on Sirius, and he quickly discovered the dark companion to the Dog Star. Subsequent studies of the companion, which was called Sirius B, or "the Pup," led to the conclusion that it was hotter than the Sun, with a diameter less than 1 percent of that of the Sun (figures 57–58).

When they were first discovered, white dwarfs presented a paradox to astronomers. If a white dwarf couldn't produce energy through nuclear fusion, how could it generate the pressure necessary to keep it from collapsing further? It didn't seem possible, yet there they were, glowing dimly and reminding scientists that "the fault is not in the stars, but in their theories," to paraphrase Shakespeare. An explanation would have to await the development of the quantum theory of matter half a century later.

A star is born when a cloud of gas and dust collapses to the point where the material in the center of the clump is so dense and hot that the nuclear fusion of hydrogen nuclei into helium nuclei can occur. The outflow of energy released by these reactions provides the pressure necessary to halt the collapse.

The nuclear reactions inside stars are nuclear fusion reactions in which the nuclei of light elements are fused together to form heavier elements (for example, hydrogen nuclei are combined to form a helium nucleus) with the release of energy. In contrast, nuclear power plants on Earth produce energy through nuclear fission, in which the nuclei of heavier elements such as uranium split apart to form smaller nuclei with the release of energy. Fusion of hydrogen into helium in the core of a star can sustain a star like the Sun for billions of years. The Sun is now in this long-lived phase of its evolution, called the main-sequence phase.

A star experiences an energy crisis, and its core collapses, when the star's basic, nonrenewable energy source—hydrogen—is used up. A shell of hydrogen on the edge of the collapsed core will be compressed and heated. The nuclear fusion of the hydrogen in the shell will produce a new surge of power that will cause the outer layers of the star to expand until it has a diameter a hundred times its present value. This is called the "red giant" phase of a star's existence.

A hundred million years after the red giant phase, all of the star's available energy resources will be used up. The exhausted red giant will puff off its outer layer, leaving behind a hot core. This hot core is called a Wolf-Rayet type star, after the astronomers who first identified these objects. This star has a surface temperature of about fifty thousand degrees Celsius and is rapidly boiling off its outer layers in a "fast" wind traveling 3.7 million miles per hour.

The radiation from the hot star heats the slowly moving red giant atmosphere and creates a complex and graceful filamentary shell called a planetary

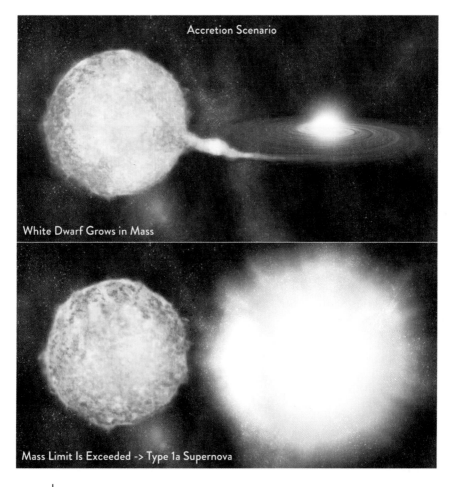

Accretion Scenario

White Dwarf Grows in Mass

Mass Limit Is Exceeded -> Type 1a Supernova

53 | RED GIANT TRIGGER MECHANISM

An artist's illustration of a red giant trigger mechanism for a Type Ia supernova. Gas pulled from a Sun-like star onto a white dwarf via a red disk can push the white dwarf over its stable weight limit and cause it to explode.

nebula (so called because it looks like the disk of a planet when viewed with a small telescope). X-ray images reveal clouds of multimillion-degree Celsius gas that have been compressed and heated by the fast stellar wind. Eventually the central star will collapse to form a white dwarf star.

In the white dwarf state, all the material contained in the star, minus the amount blown off in the red giant phase, will be packed into a volume

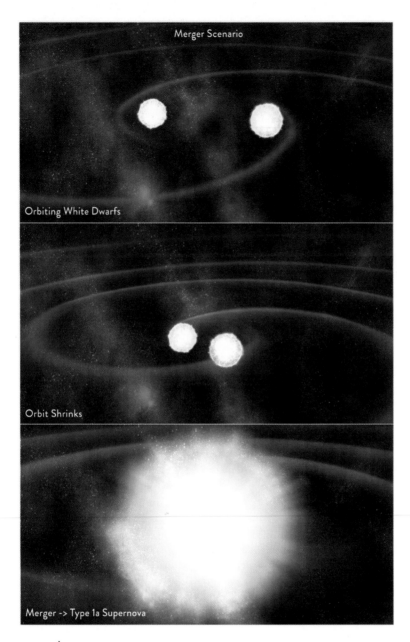

Merger Scenario

Orbiting White Dwarfs

Orbit Shrinks

Merger -> Type 1a Supernova

54 | MERGING WHITE DWARF TRIGGER MECHANISM

A trigger mechanism for a Type Ia supernova. Two white dwarf stars orbiting each other lose energy via gravitational radiation and eventually merge. If the total mass of this merger exceeds the weight limit for a white dwarf, the merged star is unstable and explodes.

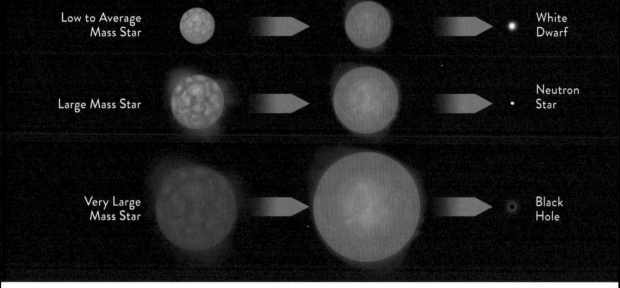

Low to Average Mass Star → White Dwarf

Large Mass Star → Neutron Star

Very Large Mass Star → Black Hole

55 | THE FATE OF STARS

A schematic (not to scale) showing that the fate of stars depends on their mass. Stars that reach the end of their evolution with a mass about 1.4 times the mass of the Sun become white dwarfs. The dividing line between the formation of neutron stars and black holes is less certain.

one-millionth the size of the original star. An object the size of an olive made of this material would have the same mass as an automobile. For a billion or so years after a star collapses to form a white dwarf, it is "white" hot, with surface temperatures of about twenty thousand degrees Celsius.

It is at this point that degenerate electron pressure kicks in to support the star. Quantum theory shows that matter in so-called degenerate states of extremely high density can produce a type of pressure called electron degeneracy pressure. This is because, according to one of the fundamental principles of the theory, no more than one electron can occupy the same energy state. To see how this works, think about a parking tower connected to a shopping mall. Only one car is allowed per space. When business is slow, the parking lot is nearly empty, and there is very little movement among the cars. From time to time, a car will enter and park on the lower level, and one will leave. As the holiday season approaches, however, the situation changes. The parking lot begins to fill up, and spaces are harder to come by. Drivers race from one level to the next searching for an empty pace. The pressure builds to get into position whenever a space opens up.

The extremely dense matter of a white dwarf is like a parking lot in a shopping mall on the last weekend before Christmas. All of the low-energy levels are degenerate or filled, so the electrons are forced into higher-energy states, not because they are being supplied with energy from another source but because there is nowhere else to go. This creates a "degenerate" electron pressure (*degenerate* refers here not to the moral character of the electrons but to the fact that all the low-energy states are occupied). This pressure is what prevents white dwarf stars from collapsing under their own weight.

In 1926, Ralph Fowler of Cambridge University, using the new theory of quantum mechanics, proposed that white dwarf stars are supported by degenerate electron pressure, a peculiar property of very dense material. A few years later, one of Fowler's research students, Subrahmanyan Chandrasekhar, known to his friends and colleagues as Chandra, worked out the theory in detail. In 1999, NASA named the Chandra X-Ray Observatory in his honor.

Chandra used Einstein's relativity theory and quantum mechanics to show that degenerate electron pressure can do only so much. If the mass of the white dwarf becomes greater than about 1.4 times the mass of the Sun—called the Chandrasekhar limit—it will collapse to become a neutron star or black hole, or it will blow itself apart in a supernova. This discovery created a big stir, since it showed that the theories of quantum mechanics and relativity could help explain cosmic mysteries. But not everyone agreed.

One of the skeptics was Arthur Eddington, a towering figure in astrophysics. He had led one of the solar eclipse expeditions that confirmed Einstein's general theory of relativity, and he had written a prescient book on stellar evolution in which he anticipated that the newly developed quantum theory would solve the question as to how stars burned. Eddington had also befriended Chandra, who as a young scientist had emigrated to England from India. But he thought that Chandra was wrong about the implication that massive stars were doomed to collapse. "I think there should be a law of Nature to prevent a star from behaving in this absurd way!" he protested (1935). The controversy stewed for several years and ended the friendship, but in the end, everyone who studied Chandra's work—everyone but Eddington, that is—agreed that Eddington was wrong and Chandra was right.

In the normal course of evolution, a white dwarf just cools down over the course of billions of years to become an inert stellar cinder. If, however, a white dwarf is a member of a binary system, and accretion of matter from a companion star pushes a white dwarf star over the Chandrasekhar limit, it becomes unstable, heats up, and explodes like a thermonuclear bomb, leaving nothing behind. The expanding cloud of ejecta glows brightly for many weeks as radioactive nickel produced in the explosion decays into cobalt and then iron.

A Type Ia supernova occurs every two hundred years or so in a galaxy the size of our Milky Way. The supernova is extremely bright, with an optical luminosity of several billion Suns for a few months. The expanding stellar ejecta disperse heavy elements that enrich the interstellar gas and create cosmic sonic booms, or shock waves, that heat the interstellar gas, accelerate particles to extremely high energies, and trigger the formation of new stars.

Interstellar matter is swept up by the supernova shock wave, often called the forward shock wave, into a thin shell. This material, which is mostly in the form of hydrogen gas, with a few impurities of heavier elements and dust grains, is heated to temperatures of millions of degrees, and broken down into a mix of protons and electrons. The shock wave also accelerates some of the electrons and protons to extremely high energies. Electrons spiraling around the magnetic field behind the shock wave produce radiation—called synchrotron radiation because it was first observed in synchrotron particle accelerators on Earth— over a wide range of wavelengths from radio waves to X-rays. Radio astronomers detected radio synchrotron radiation when they discovered the remnants of Tycho's and other supernovas.

As the supernova expands, the shell of hot, swept-up interstellar matter exerts a back pressure on the expanding stellar ejecta. This back pressure drives what is called a reverse shock wave back into the stellar ejecta. We encounter reverse waves—hopefully not shock waves—frequently when driving in rush hour traffic on the highway. A vehicle far ahead of us may slam on the brakes to avoid a road hazard. The next vehicle in the lane will brake to avoid a collision, and so on, sending a wave of braking and brake lights moving back down the lane. In a similar way, the reverse shock wave sweeps through the stellar ejecta and heats it up to millions of degrees.

With Chandra, it has become possible to observe both the shock wave and the ejecta heated by the reverse shock wave. The Chandra image of the Tycho supernova remnant (figure 52, page 158) reveals fascinating details of the turbulent debris created by the supernova. The forward shock wave produced by the expanding debris is outlined by the strikingly sharp blue circular arcs seen on the outer rim. The stellar debris, which has a temperature of about 10 million degrees and is visible only in X-rays, shows up as mottled yellow, green, and red turbulent clumps of gas. No central point source is detected. The absence of such a source is consistent with other evidence that Tycho's supernova signaled the

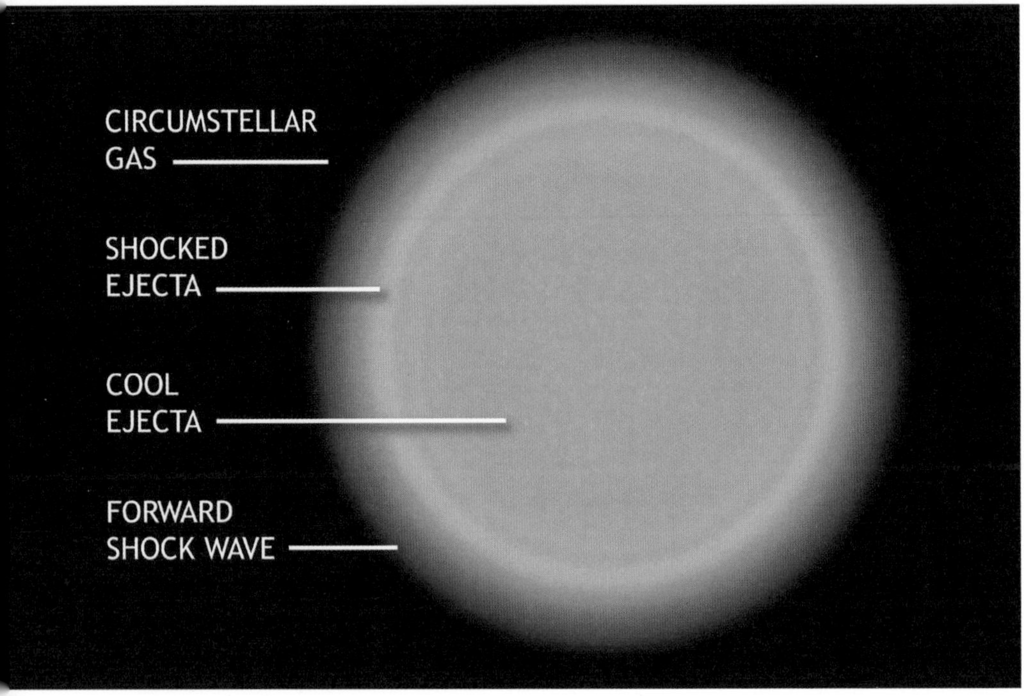

CIRCUMSTELLAR
GAS

SHOCKED
EJECTA

COOL
EJECTA

FORWARD
SHOCK WAVE

56 | SUPERNOVA EXPLOSION

A supernova explosion of a star, which ejects most (Type II), and in some cases all (Type Ia), of the star into the circumstellar gas at millions of miles per hour. This event generates shock waves that produce shells of hot gas and high-energy particles that are observable as supernova remnants for hundreds to thousands of years after the explosion. A forward shock wave speeds ahead of the ejecta and, like an extreme version of the sonic booms produced by airplanes' supersonic motion, produces sudden large changes in pressure and temperature behind the shock wave (violet). The hot high-pressure gas (purple) behind the forward shock expands and pushes back on the ejecta, causing a reverse shock wave that heats the ejecta (orange). Eventually the reverse shock wave traverses the cool ejecta (blue) and heats it. An observer surfing along with the front edge of the ejecta would see the reverse shock moving inward and the forward shock moving outward. A distant observer would see both shells moving outward at differing velocities.

57 | OPTICAL IMAGE OF SIRIUS A AND B STAR SYSTEM

Sirius A (the Dog Star), only 8.7 light-years from Earth, is the brightest star in the northern sky when viewed with an optical telescope, whereas Sirius B is ten thousand times dimmer. The two stars are so close together that Sirius B, a white dwarf, escaped detection until 1862, when Alvan Clark discovered it. Sirius B's mass is equal to that of the Sun, but it is stuffed into a diameter only 90 percent that of Earth. Gravity on the surface of Sirius B is four hundred thousand times that on Earth.

detonation and destruction of a white dwarf star. Hardly anyone argues with this conclusion. Exactly how the white dwarf explodes in a Type Ia supernova, though, has been the subject of wild swings of opinion.

Is a Type Ia supernova caused by a death spiral of two white dwarfs that merge and then explode (the double-degenerate model), or is the explosion caused by a white dwarf that accretes so much matter from a red giant companion star that it is pushed over the Chandrasekhar limit (the single degenerate model)? Both scenarios have been around for more than three decades, and both have gone in and out of favor. In a 2001 review, the double degenerate model was all but consigned to the trash heap of failed models, but by 2012, the situation had been reversed, partly due to improved theoretical modeling of the explosion process, and it was cited as the most promising model. Tycho's supernova is a poster child for this puzzle, with some research papers saying that researchers have found evidence for the surviving companion star of the supernova, whereas others report that they have done exhaustive searches and come up empty.

The specific characteristics of the progenitor binary system are expected to have an important imprint on the circumstellar medium into which the supernova shock ejecta expand. In particular, if one of the companion stars is a red giant, the wind usually associated with a red giant would modify the environment of the pre-supernova system by creating a low-density cavity bounded by a shell of gas. This modification would have detectable effects on the evolution of the shock wave, which should be detectable in the shape of the supernova remnants and their spectra. Chandra observations of a sample of seven Type Ia supernova remnants show no evidence for a fast wind, contrary to expectations for some single degenerate models.

Then there is the remnant of the supernova named after Johannes Kepler, which does show evidence of a fast wind. Kepler was not the first to observe the supernova, but he made it "his" by observing it continually for a year, beginning shortly after it had appeared, and followed this up with the publication of a book describing his observations in great detail. The Chandra image of the Kepler supernova remnant (figure 59) reveals a rapidly expanding nebula that shares features with the image of Tycho's supernova remnant: a thin blue shell defining the high-energy electrons produced by the supernova shock wave, and the turbulent interior of the shocked stellar debris. However, Kepler's supernova remnant shows more structure than the chaotic interior of Tycho's supernova remnant.

The bright northern rim of Kepler's supernova remnant has been interpreted as an enhancement caused by the shock wave running into material

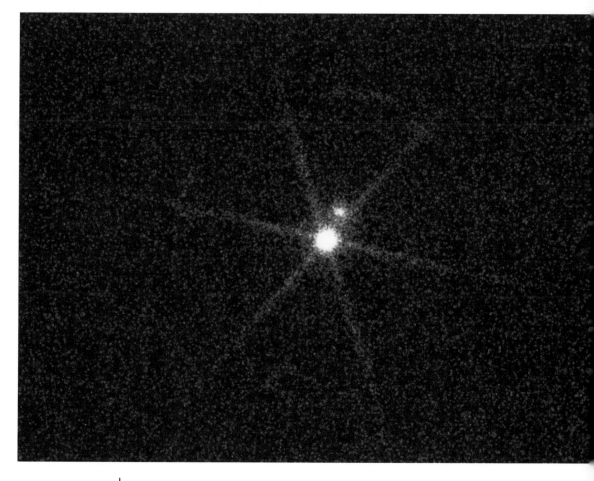

58 | X-RAY IMAGE OF SIRIUS A AND B STAR SYSTEM

A Chandra image showing two X-ray sources. In contrast to figure 57, an optical image of the system, the central, brighter source in this X-ray image is Sirius B, the white dwarf, with a surface temperature of about twenty-five thousand degrees Celsius. Although Sirius B produces very low-energy X-rays, its total power output in X-rays is much greater than that of Sirius A. The small source at roughly one o'clock is Sirius A—a normal blue-white star more than twice as massive as the Sun. X-radiation from Sirius A may be even weaker than indicated here, as the dim source at its position may be due to ultraviolet radiation from Sirius A leaking through the filter on Chandra's detector. The spike pattern in this image is caused by the support structure of Chandra's transmission grating.

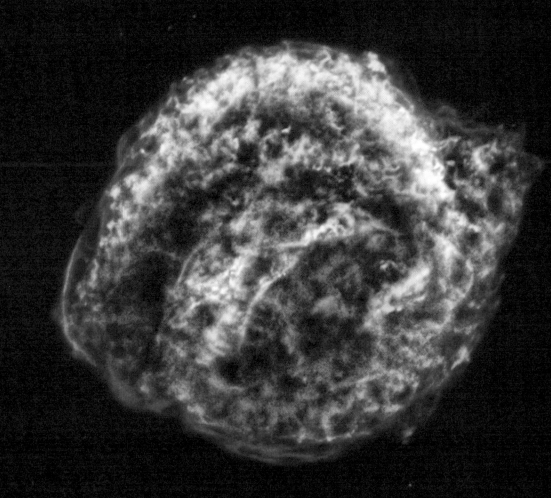

59 | KEPLER SUPERNOVA REMNANT

A remnant about twenty thousand light-years from Earth. The bright blue outer rim is due to high-energy X-rays from the forward shock wave produced by the explosion. Low- and intermediate-energy X-rays from multimillion-degree gas heated by the reverse shock wave are red and green, respectively. The bright arc at the top has been interpreted as a bow shock produced by the remnant's motion through the progenitor star's red giant wind. The disk-shaped structure near the center could have been caused by supernova debris colliding with a disk of material that the giant star expelled before the explosion.

ejected before the supernova by a red giant companion to the white dwarf star. In addition, a disk-shaped structure near the center of the remnant is apparent, and it provides two more pieces of evidence for a giant companion star: an enhanced amount of magnesium, an element not produced in Type Ia supernovas in any great abundance; and bright infrared emission, also indicative of a preexisting dusty circumstellar disk.

Theoretical simulations seeking to reproduce the outer rim indicate that the progenitor was a binary star system consisting of a white dwarf and a giant companion star of four to five solar masses. In this case, the white dwarf exploded only about 100 million years after it had formed.

If such "prompt" supernovas occur, they could have implications far beyond the galaxy. Because Type Ia supernovas all occur in stars that have a mass of about 1.4 solar masses, they produce about the same amount of light and have a characteristic light curve. As discussed earlier, this property makes them extremely useful as distance indicators. If one Type Ia supernova is dimmer than another, it must be farther away by an amount that can be calculated. In recent years, Type Ia supernovas have been used to determine the rate of expansion of the universe. This research has led to the discovery that the universe's expansion is accelerating, possibly because it is filled with a mysterious substance called dark energy. One constraint on this method is that it takes time, roughly a billion years, for an appreciable number of the first round of Type Ia supernovas to occur after the Big Bang. If there are enough of these prompt Type Ia supernovas, it may be possible to measure the effects of dark energy on cosmic expansion during earlier times.

As we learn more about them, other Type Ia supernovas continue to provide surprises. Chandra's observations of the youngest known (about 110 years old) galactic supernova remnant, G1.9+0.3, have yielded the first definite detection of radiation from radioactive decay of scandium nuclei, which in turn came from the radioactive decay of titanium. From the strength of the scandium X-radiation, it is possible to compute the amount of titanium produced in the explosion: a little more than 7 million times the amount of titanium in Earth's crust, enough to make a fair large number of airplanes, golf clubs, and hip joints. Theoretical analysis of how so much titanium could have been produced seems to require a lopsided explosion, so that the necessary shock waves can propagate faster, heat more, and synthesize more titanium. The spatial distribution of iron and the intermediate-mass elements silicon and sulfur suggests that the explosion itself was asymmetric. Incidentally, the estimated amount of iron produced in the explosion is roughly equivalent to the iron in a million Earths.

60 | SUPERNOVA 1987A

A supernova in the Large Magellanic Cloud, a galaxy 160,000 light-years from Earth. SN 1987A, which was visible to human eyes, is the brightest known supernova in almost four hundred years. This composite optical (Hubble; pink, white) and X-ray (Chandra; blue, purple) image shows the triple ring structure, with the bright equatorial ring. Brilliant optical hotspots now encircle the ring; these are multimillion-degree gas shown by Chandra's data. The image is about fifty light-years on each side.

17

Things fall apart; the centre cannot hold;
Mere anarchy is loosed upon the world.

William Butler Yeats, "The Second Coming," 1919

CORE
COLLAPSE

Sun-like stars shine for billions of years. In contrast, massive stars lead short, spectacular lives. After only a few million years, a star that is a dozen or more times as massive as the Sun is using energy prodigiously and rushing headlong toward catastrophe. First, the massive star expands enormously to become a red giant and ejects its outer layers at a speed of about twenty thousand miles per hour. A few hundred thousand years later—a blink of the eye in the life of a Sun-like star—the intense radiation from the exposed hot inner layer of the massive star begins to push gas away at speeds in excess of 3 million miles per hour.

After the onset of the fast stellar wind phase, a massive star is doomed. In the next few hundred thousand years, the interior of the star takes on a layered structure, with the products of successive thermonuclear fusion episodes in different layers (figure 61). From the inside out, there is a group of heavy elements such as iron, nickel, chromium, and titanium; then sulfur, silicon, magnesium, neon,

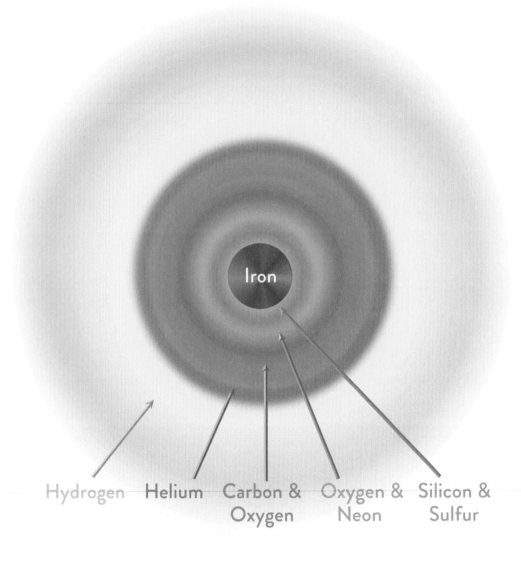

Hydrogen   Helium   Carbon &   Oxygen &   Silicon &
                      Oxygen   Neon   Sulfur

61 | STAR BEFORE SUPERNOVA

The structure of a star just before it goes supernova. As the star nears the end of its evolution, heavy elements produced by nuclear fusion inside it are concentrated toward the center.

oxygen, carbon, and helium; and finally unprocessed hydrogen-rich material. The nuclear reactions creating this onionskin structure start slowly, taking millions of years to fuse hydrogen to helium, but each phase proceeds more rapidly until the creation of the iron group in the core, which takes only about a day. This is the beginning of the end because no energy is generated from the nuclear fusion of iron.

With no energy being created in the core, the center cannot hold, and it collapses in a free-fall. In a few seconds, it is all over. Matter in the core is crushed to extreme densities. Electron degeneracy pressure cannot save the day, as electrons are pushed inside protons to form one of the strangest objects in nature: a neutron star, with a mass equal to the Sun, which is about three hundred thousand times the mass of Earth, is crammed into a region a little more than twelve miles across (figure 62).

Neutron star matter is no ordinary type of matter. The material that we, and everything around us, are made of consists largely of empty space. Even a rock is mostly empty space. This is because matter is made of atoms—nuclei composed of protons and neutrons surrounded by orbiting clouds of electrons. The nucleus contains more than 99.9 percent of the mass of an atom, yet it has a diameter of only one one hundred thousandth that of the electron cloud. The electrons themselves take up little space, but the pattern of their orbit defines the size of the atom, which is therefore 99.9999999999999 percent open space. When we bump into a rock, it feels painfully solid. Yet it is mostly empty space, a hurly-burly of electrons moving so fast that we cannot see—or feel—the emptiness. In the prelude to the formation of a neutron star, gravitational forces become so strong that all this emptiness is crushed out of the atoms as electrons are forced into protons.

What would matter look like if it weren't empty, if we could crush the electron cloud in each atom down to the size of its nucleus? If we could generate a force strong enough to crush all the emptiness out of a rock roughly the size of a football stadium, the rock would be squeezed down to the size of a grain of sand, but it would still weigh 4 million tons. In a neutron star, all the emptiness has been squeezed out. A new kind of pressure, neutron degeneracy pressure, kicks in. It is similar to electron degeneracy pressure, except now the pressure is provided by neutrons rather than electrons, and the diameter of a neutron star is about a thousand times smaller than that of a white dwarf.

What happens after the creation of a neutron core in the center of a star is still the subject of considerable debate, but there is general agreement that

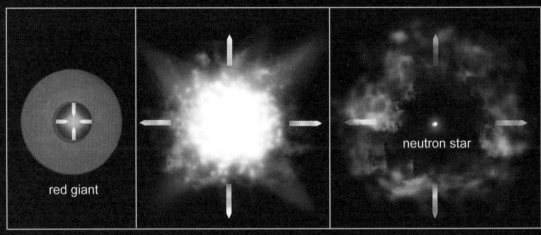

Birth of a Neutron Star and Supernova Remnant
(not to scale)

red giant

neutron star

Core Implosion ➡ Supernova Explosion ➡ Supernova Remnant

62 | NEUTRON STAR AND SUPERNOVA REMNANT

The exhaustion of the nuclear power source at the center, or core, of a massive star, leading to core collapse. In less than a second, a neutron star (or a black hole, if the star is extremely massive) is formed. This formation releases an enormous amount of energy in neutrinos and heat, which reverses the implosion. Everything but the central neutron star is blown away at speeds in excess of 30 million miles per hour as a thermonuclear shock wave races through the now expanding stellar debris, fusing lighter elements into heavier ones and producing a brilliant visual outburst that can be as intense as the light of several billion Suns.

it is not good for the rest of the star. The formation of the neutron star releases enormous amounts of energy, creating a titanic shock wave. This shock wave sweeps through the remaining outer layers of the star, fusing lighter elements into heavier ones. The result is a supernova, which shines with the power of several hundred million Suns for several months. Unlike the case of the Type Ia explosion, which destroys a white dwarf, the neutron star survives. Neutron stars, with their extreme gravitational fields, extreme density, and extreme electromagnetic fields, provide a valuable cosmic laboratory for the study of matter under conditions that cannot be reproduced on Earth or in most other places in the cosmos, for that matter.

As the supernova ejecta expand, they create a shock wave in the surrounding gas. The shock wave sweeps through the stellar wind that flowed away from the star in the millions of years before it exploded. The details of this shock wave—its speed, the intensity and spectrum of the X-ray emission—depend on the properties of the stellar wind, which in turn depend on the structure and energetics of the star. Scientists have used Chandra to study the X-ray emission from supernovas that are a few years to a few decades old and thus gain insight into the pre-supernova existence of the stars.

In 1987, a core-collapse supernova, called Supernova 1987A, or SN 1987A for short, occurred in the nearby Large Magellanic Cloud galaxy (figure 60, page 176). SN 1987A has given astrophysicists a front-row seat (well, at least a seat in the front row of the balcony—we wouldn't want to be much closer) to one of nature's grandest spectacles. The detection of neutrinos and the observation of the expansion of the ejecta confirmed our basic ideas of how a core-collapse supernova occurs. As usual in such cases, though, many questions remain.

The type of star that exploded is known with certainty because it is no longer there. It was a blue supergiant star. The rest of the story is complicated, and it illustrates the difficulties of stellar forensics. A blue supergiant is expected to have gone through the red giant phase and produced a slow wind, followed by the blue supergiant phase, when a fast wind rear-ended the red supergiant wind and carved out a hot, low-density cavity. What in fact was observed with optical telescopes was a striking triple ring system. This structure might be explained by the merger of two stars about twenty thousand years before the explosion. In this picture, the angular momentum of the two stars would have caused the slow red supergiant wind to take on an hourglass shape, and the collision of the fast wind would have created the observed set of rings. Another possibility is that the triple ring structure was produced by loss of mass from a rapidly rotating star. In both cases, an equatorial ring of gas was produced and the SN 1987A blast wave swept through that ring, producing the bright waist of the hourglass.

Chandra has been observing SN 1987A for sixteen years, starting with the launch of the telescope about twelve years after the supernova and continuing to the present. The Chandra observations trace the progress of the shock wave and probe the gas around the star. About sixteen years after the explosion, the shock wave moved through and lit up the dense equatorial ring of gas created by the collision of the fast and slow winds long before the explosion. At twenty-six years, in mid-2013, the X-ray intensity had flattened and remained approximately constant. The shock wave apparently had left the dense material of the known equatorial ring and moved into the unknown territory beyond. One of the best indications of what lies beyond will come from Chandra's observations of the X-ray signal produced as the shock wave expands into the interstellar space around SN 1987A.

Meanwhile, astrophysicists are eagerly awaiting the appearance of the neutron star that is expected to exist in the central regions of SN 1987A. They may have to be patient. The neutron star has yet to be detected by Chandra, which suggests that it is still obscured by surrounding dust and gas. It may be another hundred years before that veil lifts.

In one well-studied supernova remnant, the veil has already lifted. Although it has been more than seventeen years, I remember clearly the moment it was first seen. Chandra had been launched and deployed, and all systems were go to create the first significant image at the focus of a history-making X-ray telescope. Cassiopeia A (Cas A; figure 51, pages 156–57), the remnant of a core-collapse supernova, would be the target. Cas A is a strong source of X-rays and has been observed by every previous X-ray telescope. The Chandra image of Cas A would provide an immediate and demanding test of just how good this new telescope was.

Scientists and engineers who had worked on the telescope, most for almost a decade and some for more than two, crowded into the instrument room, where the image would be displayed on a single computer monitor. Then suddenly, at 8:40 p.m. on August 26, 1999, there it was: a gorgeous image of the remains of a star that had exploded ten thousand light-years away. We were looking at the best X-ray image ever made of an object outside our solar system. It was as if we were looking over Galileo's shoulder, peering through the eyepiece of his telescope, and getting a view of the universe that no one had ever seen before. Words are inadequate to describe the feeling, but people tried.

"Awesome!"

"Spectacular!"

"What's that in the middle?"

What indeed? A tiny bright dot was visible in the middle of the remnant. In its first look, Chandra had discovered what no previous telescope could find: the neutron star left over from the explosion that had produced Cas A three and a half centuries earlier.

Neutron stars are formed at temperatures of billions of degrees, and they cool rapidly at first, then more slowly. Exactly how a neutron star cools depends on what is going on in the interior of the star, and therein exists an opportunity. Physicists who specialize in the behavior of matter at extremely high densities make calculations based on hypothetical models and compare them with the observed cooling rate. The strangeness of neutron stars begins in the atmosphere. The Cas A atmosphere is only four inches thick and is composed of almost pure carbon. It has a pressure more than ten times that at the center of Earth, and a density similar to diamonds. The gravity at the surface of a neutron star is 100 billion times stronger than at the surface of Earth, and the temperature is about 1.5 million degrees Celsius.

Chandra archival data show that the X-ray intensity of the Cas A neutron star has declined by 21 percent over a ten-year period. When combined with the model for the carbon atmosphere, this implies that the surface temperature has declined by 4 percent over this time. This is faster than standard models predict, which may mean that the interior is in a superfluid or superconducting state. Under these conditions, heat can flow almost without resistance from the interior of the star to the surface. On Earth, superconductors, or superfluids, are materials cooled to near absolute zero (−273 degrees Celsius) in which electrical currents or fluid flow occur without resistance. The ability of matter with temperatures in excess of a million degrees to exhibit superfluidity and superconductivity shows that when considering neutron stars, we have entered a realm where the extraordinary has become ordinary.

The Cas A neutron star is embedded in the rapidly expanding debris from the supernova explosion. Since Chandra's initial observation, Cas A has been observed repeatedly. Various lines of evidence, including light echoes from the original event, suggest that Cas A is the product of a core-collapse supernova. Evidence also shows that the pre-supernova star started out with a mass of about sixteen Suns and shed two-thirds of this mass during the course of its evolution. Chandra data from accumulated deep exposures of Cas A reveal two large, opposed jetlike structures extending out to about ten light-years from the center

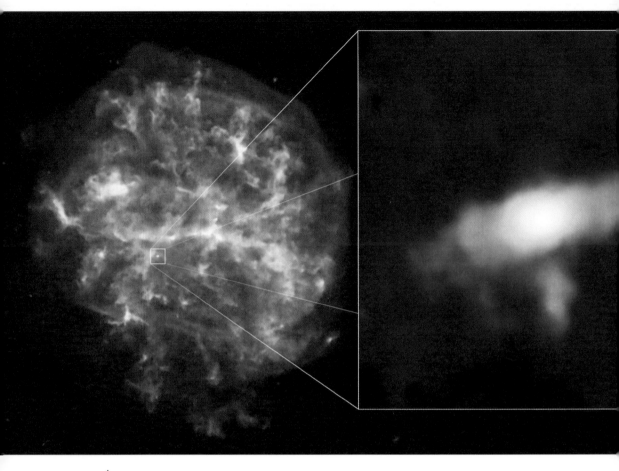

63 | G292.0+1.8 SUPERNOVA REMNANT

The remnant of a core-collapse supernova twenty thousand light-years from Earth. Chandra's image shows red, orange, green, and blue that represent low-, medium-, medium high-, and high-energy X-rays, respectively. Optical radiation (from the Sloan Digitized Sky Survey) is white. The bright horizontal belt across the middle of the remnant is due to shocked circumstellar material, and the pullout shows the neutron star as a white spot in a blue-purple cloud—the pulsar wind nebula. The main image is about sixty-five light-years across.

of the remnant. To give some perspective, note that the bright star Sirius is only 8.6 light-years from Earth, so these jets extend well into interstellar space. Clouds of iron that have remained nearly pure for the approximately 340 years since the explosion were also detected. The offset of the neutron star—the white dot in the center of the Chandra image of Cas A—is real and likely was produced by recoil from an asymmetric explosion.

A three-dimensional model of the Cas A supernova remnant has been constructed from observations with Chandra data and infrared Spitzer data. Jets composed mostly of silicon appear in the upper left and lower right, whereas plumes of iron are seen in the lower left and top. The 3D reconstruction indicates that the jets and plumes emerge from a broad, thick disk, whereas shock waves from the explosion have puffed up the remaining gas into a roughly spherical shape. The forward shock wave is not shown in the reconstruction.

The yellow, orange, and green material is debris from the explosion that has been heated by the reverse shock wave. The red material inside of the yellow/orange ring has not yet encountered the inward-moving shock. It cooled rapidly as the ejecta expanded—the same process that cools aerosol spray from a can—so it glows only in infrared light. The implication of this work is that the outer layers of the star are ejected spherically, but the inner layers come out in a disk, with high-velocity jets spurting out in multiple directions. This means that the explosion of a star is a complicated affair, likely related to the rotation of the star or its membership in a binary star system, or both.

Another supernova remnant, G292.0+1.8 (figure 63), illustrates many aspects of the evolution of the remnant of a core-collapse supernova: a blast wave moving through previously ejected circumstellar material, ejecta enriched with heavy elements, and a central neutron star. It also exhibits an additional feature not present in Cas A: a pulsar wind nebula created by the neutron star, which is highly magnetized and rotating rapidly. The offset of the neutron star from the center of G292.0+1.8 may be due to the recoil from an asymmetric explosion, as in Cas A. The progenitor likely had a mass in the range of twenty to thirty-five Suns. The detection of the neutron star and the pulsar wind nebula conclusively associates this supernova remnant with the core collapse of a massive star. One of the major advances of Chandra has been the detection of numerous pulsars and their associated pulsar wind nebulas. These nebulas are proving to be one of the best ways to identify supernova remnants produced by the core collapse of a massive star. Before Chandra, it was difficult to tell if a supernova remnant was the product of a core collapse or the thermonuclear disruption of a white dwarf star.

Out of the complexity and diversity of the remnants of core-collapse super-novas, some general similarities seem to be emerging. They come from stars that originally had a mass in the range of fifteen to thirty-five Suns. The remnants are very irregular and show large deviations from spherical symmetry, with evidence of bipolar structure and jets. The origin of these structures is unknown. They could be due to a companion star and accretion disk, though none has been found, or to instabilities in the explosion process, or to rapid rotation, or to strong magnetic fields, or to a combination of all of the above.

The fate of the cores of the super-heavyweight stars, with masses above about thirty Suns, is uncertain. They could stabilize as neutron stars and eject massive envelopes, or they could collapse to black holes with or without the ejection of envelopes of gas. Compelling evidence that Cygnus X-1 and about a dozen other binary X-ray sources contain black holes shows that, unless black holes form directly without an explosion—considered possible but not likely—some supernova remnants must contain black holes. One of the best candidates for a supernova remnant that contains a black hole is W49B (figure 64).

The shape of W49B indicates that it is the product of an explosion in which matter was ejected at high speeds along the poles of a rotating star. This is in keeping with some models of supernovas in rapidly rotating stars. Further evidence comes from the distribution of iron in the supernova remnant. The jet shows an enhanced abundance of iron, which could have come from the radioactive decay of nickel produced in the explosion. This is consistent with the predicted yields of nickel in models of bipolar/jet-driven supernovas. In these events, the kinetic energy of the expanding matter is greater at the polar axis, and the thermonuclear reactions that produce nickel are more efficient. An intensive search of Chandra data for evidence of a neutron star in W49B has come up empty, suggesting that the supernova left behind a black hole. The estimated age of W49B is one thousand years. If W49B does contain a black hole, it would be the youngest known black hole.

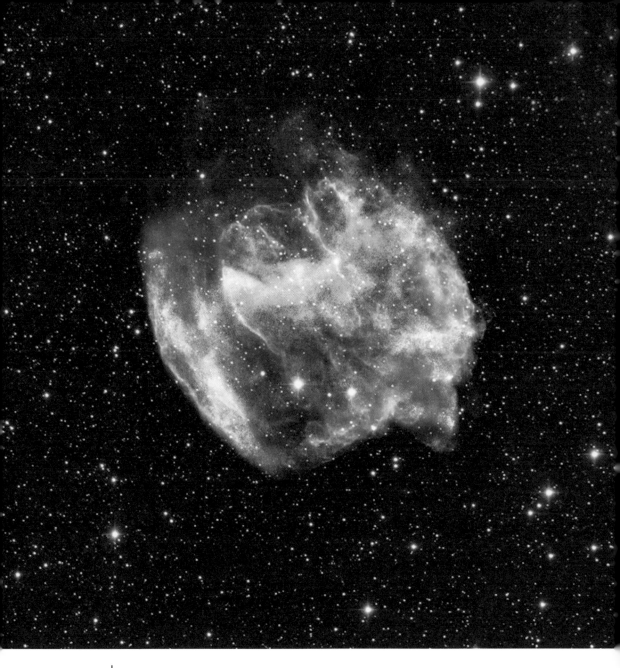

64 | W49B SUPERNOVA REMNANT

A highly distorted remnant that may contain the most recent black hole formed in the Milky Way. The image combines Chandra X-ray data (blue and green), Jansky Very Large Array radio data (pink), and Palomar Observatory infrared data (yellow).

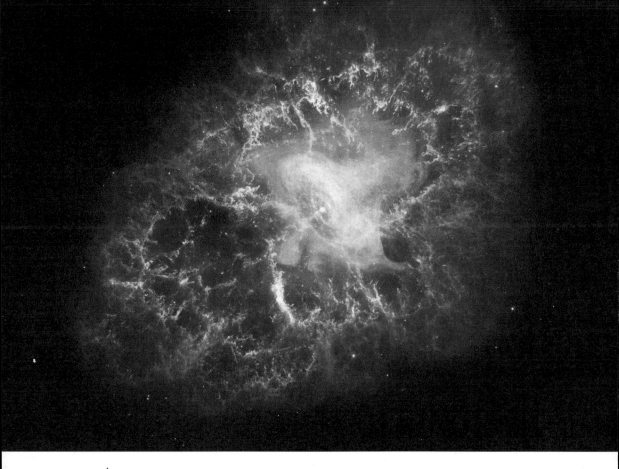

65 | CRAB NEBULA

A composite image of the Crab Nebula at infrared, optical, and X-ray wavelengths. In 1054 CE, Chinese astronomers and others around the world noticed a new bright object in the sky. This "new star" was, in fact, the supernova explosion that created what is now called the Crab Nebula, in the constellation Taurus. At the nebula's center is an extremely dense, rapidly rotating neutron star left behind by the explosion. This star, also known as a pulsar, emits a storm of high-energy particles. Chandra X-ray data is shown in blue (around the neutron star), the Hubble optical image is in red and yellow, and the Spitzer infrared image is in purple. Super-energetic electrons emitting X-rays radiate away energy more quickly than lower-energy electrons emitting optical and infrared light, and so the blue Chandra area is smaller than the others.

18

It seems, therefore, that when the oblique
rotator model is realized, it can
lead to a release of energy from a
neutron star.

Franco Pacini, "Energy Emission from a Neutron Star," 1967

THE CRAB AND OTHER
PULSAR WIND NEBULAS

Theoretical models for the evolution of a massive star show that in the last
phases of the star's evolution, the core collapses to form an ultradense object
called a neutron star. What happens next is still not well understood, but there is
general agreement that the formation of the neutron star releases a tremendous
amount of energy, resulting in the ejection of the outer layers of the star at high
speeds in a supernova explosion.

In a companion paper to their 1934 paper describing a new class of cosmic explosions, which they called "supernovae," Walter Baade and Fritz Zwicky
wrote: "With all reserve we advance the view that a super-nova represents the
transition of an ordinary star into a neutron star, consisting mainly of neutrons." This hypothesis was made only two years after James Chadwick had discovered neutrons in his laboratory in Cambridge, England. It would be another
thirty-four years before neutron stars would be discovered. By the 1960s,

theoretical physicists had accepted their existence and were impatiently waiting for astrophysicists to find the observational proof.

Actually, a neutron star had been identified twenty years earlier, but no one realized it. In 1942, Nicholas Mayall of Lick Observatory in California and Jan Oort of Leiden Observatory in Holland studied the Chinese and Japanese historical records of the 1054 CE event and concluded, "The Crab Nebula may be identified with the 1054 supernova." Rudolph Minkowski wrote a paper shortly after the work by Mayall and Oort in which he correctly identified a star in the center of the Crab Nebula (figures 65–66) as the remnant of the explosion. He erroneously concluded, however, that the central star is a white dwarf.

As an aside, the Crab Nebula got its name from a sketch made by William Parsons, a.k.a. Lord Rosse, an Anglo-Irish astronomer who in the 1840s had a seventy-two-inch reflector built at Birr Castle in Ireland. It was the largest telescope in the world at the time. Before that, Parsons had observed a nebula with a thirty-six-inch telescope. It looked like a horseshoe crab to him, so he called it the Crab Nebula. Later observations with his larger telescope showed that it didn't really look like a crab, but the name stuck.

There the matter stood until the late 1950s, when the field of nuclear astrophysics developed. Alistair Cameron, then at Atomic Energy of Canada, updated earlier work done in the 1930s by Robert Oppenheimer and George Volkov on the properties of neutron stars. This effort exploded with the discovery of the first extrasolar X-ray sources in 1962 by Riccardo Giacconi and colleagues. A flurry of theoretical papers followed, showing that the X-ray sources could be neutron stars. In particular, it was claimed that the X-ray emission from the Crab Nebula, detected on a rocket flight in 1963, could be explained by a hot neutron star. In that case, the X-ray emission from the Crab Nebula would not be extended like that of the nebula itself. Rather, it would be pointlike, since it would be coming from a neutron star. This seemed like a safe prediction for a while, because X-ray detectors in those days were basically just boxes with some X-ray-detecting electronics and had poor angular resolution similar to that obtained by looking through a shoe box with a pinhole in it.

In spite of this obstacle, Herbert Friedman of the Naval Research Laboratory, whose team had discovered the Crab Nebula X-ray source, took up the challenge of measuring the size of the Crab Nebula X-ray source. He timed a rocket flight to coincide with a lunar eclipse of the Crab Nebula. What Friedman and his team observed was that as the Moon gradually occulted the Crab Nebula and moved across the central star thought to be the X-ray source, the X-ray emission declined gradually, not suddenly as would be expected in the eclipse of a point source. This led to the conclusion that the X-rays are coming from a region several light-years across, not a neutron star. As it turned out, that conclusion was only partially correct. We now know that, although the neutron star does not produce all of the X-rays from the Crab Nebula, it does produce a fair amount. It is also indirectly responsible for producing all of them.

The observation of Friedman's group presented a problem. The electrons that produce X-rays by the synchrotron process, the same process that produces the radio waves from the Crab Nebula, lose their energy in just a few years at most. Either electrons in the Crab Nebula, which is several light-years in extent, are being continually accelerated to extremely high energies or the neutron star is continually producing the high-energy, X-ray-emitting electrons, which then stream out into the nebula.

In 1967, Franco Pacini, who was at Cornell University at the time, showed how the high-energy electrons could be produced. A neutron star forms from the collapse of the core of a star. The size of the core is comparable to the size of Earth, maybe a little larger. As the core collapses, its rotation rate will increase, just as the rotation rate of a figure skater increases when she pulls in her arms. The magnetic field threading the core will also be greatly amplified as a consequence of the collapse. Pacini concluded that the collapse of the core of a massive star could lead to a highly magnetized, rapidly rotating neutron star. This neutron star would act like a generator, and could in principle produce a prodigious amount of energy in the form of very low-frequency electromagnetic waves. This process, Pacini speculated, could be the source of energy needed to explain the acceleration of electrons to the extremely high energies in the Crab Nebula.

Pacini's paper was published in 1967, before there was any evidence for the existence of neutron stars. The next year, quite independently, a paper describing a remarkable new type of radio source was published by Anthony Hewish, Jocelyn Bell, and their colleagues from Cambridge University. Several sources of radio waves were found to be pulsing with great rapidity and astonishing regularity. They blinked on and off with the precision of a clock that loses only one second in a million years. The time between pulses varies from one object to the next but is typically one second or less. The next year, the central star of the Crab Nebula was discovered to be pulsar at radio, optical, and X-ray wavelengths. The regularity of the pulses can be understood only in terms of the rotation of a very massive body, that is, a star. The rapidity of the pulsation, especially in the case

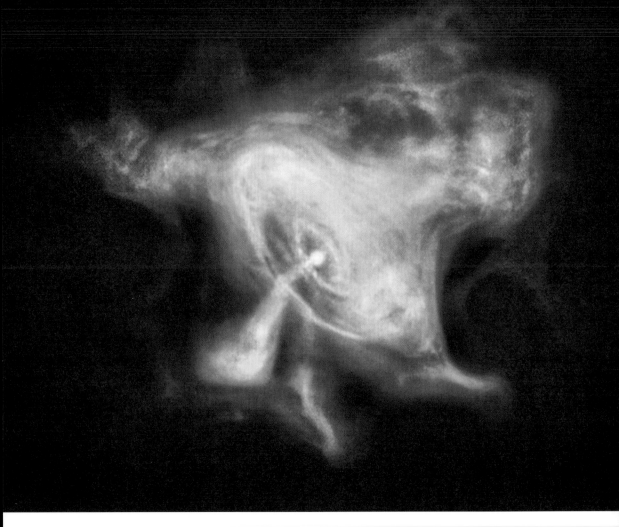

66 | CRAB NEBULA

An X-ray image of the Crab Nebula (about 6,500 light-years from Earth), the remnant of a supernova observed in 1054 CE by Chinese astronomers and others around the world. An extremely dense, rapidly rotating neutron star, also known as a pulsar, was left behind by the explosion. It produces a blizzard of high-energy particles that has created an expanding X-ray nebula. In this Chandra image, lower-energy X-rays are red, medium-energy X-rays are green, and the highest-energy X-rays are blue. The gradual transition from high-energy X-rays in the central regions to lower-energy ones in the outskirts occurs because the electrons producing the X-rays lose energy as they travel out into the nebula.

of the Crab pulsar, can be explained by only one type of star—a neutron star. Any other type of star, even a white dwarf, would break apart if it were to rotate as fast as some pulsars are rotating.

The discovery of pulsars was the key that unlocked the door to the exploration of the strange and wonderful worlds of neutron stars and black holes. The Crab Nebula contains the most energetic pulsar known. It has a pulsation period of about one-thirtieth of a second and produces, through the mechanism Pacini outlined, as much energy per second as ten thousand or more Suns.

The combination of rapid rotation and strong magnetic field produces very low-frequency electromagnetic waves, as Pacini had predicted. Another way to look at it is that the rapid rotation and strong magnetic field create electrical potentials of trillions of volts—tens of millions of times the voltages of lightning bolts. These voltages act to accelerate electrons away from the neutron star at high energies. Some of the electrons—they are likely a mix of electrons and antielectrons, or positrons, created by the enormous voltages, but I will just call them electrons—stream out along the magnetic poles of the star, producing a lighthouse-like beam of radiation. Because of the rotation of the neutron star, the beam shows up as a regular sequence of pulses.

The discovery of the pulsar in the center of the Crab Nebula solved the problem of the energy source for the X-ray emission overnight. A highly magnetized neutron star was in the center of the nebula, spinning thirty times a second and producing high-energy electrons in the process.

The energy of the electrons comes ultimately from the rotational energy of the neutron star, so the neutron star should be slowing down. A clear prediction of the model is that the amount of slowdown should be sufficient to account for the energy lost in the form of radiation, and then some, since the process is unlikely to be 100 percent efficient. Measurements of the slowing down of the pulsar's rotation rate showed that this is indeed the case. The pulses are very gradually getting farther apart, implying that the neutron star is slowly spinning down. Each successive spin takes about a trillionth of a second longer. Put another way, the pulsar clock loses about fifteen microseconds a year. Either way, the spin-down energy loss of the neutron star converts to an energy loss adequate to explain the observed radiation from the Crab Nebula.

Within a couple of years, others had further developed the model of the highly magnetized, rapidly rotating neutron star, and it became the accepted explanation for pulsars. This model received dramatic confirmation more than three decades later when Chandra made spectacular X-ray images of the region surrounding the Crab pulsar.

Case closed? In terms of the prime suspect, yes. In terms of the means, sort of. It took another quarter of a century, until the launch of Chandra, to reveal the details as to how the pulsar dynamo actually works. The central pulsar appears as a bright point source of X-rays, surrounded by a tilted ring, or torus, which is about three-tenths of a light-year from the pulsar. For comparison, the diameter of the ring is about five hundred times the size of the orbit of Pluto, so the solar system would fit comfortably inside the ring. The ring is likely a shock wave produced by the interaction between the outflowing pulsar wind and the surrounding material in the nebula. Particles flow or diffuse away from the shock wave in a predominantly disklike geometry to form a larger ring or torus farther out in the nebula. The filamentary structure seen at optical wavelengths is likely a netlike shell of cooler material surrounding the cloud of high-energy particles in the interior. A strong jet extends in a direction perpendicular to the ring, and a fainter one extends in the opposite direction.

Since 2011, intense gamma ray flares from the Crab have been reported on a regular basis. The gamma ray intensity at photon energies roughly one thousand times those observed by Chandra increases sharply in a matter of hours, then declines. The rapidity of the rise and fall implies that the flares are very likely coming from a small region about the size of our solar system. Gamma ray telescopes cannot resolve the exact location of the flares, so they could be coming from anywhere in the nebula, but the most likely location is somewhere near the inner ring, or perhaps close to the neutron star.

For now, the origin of the gamma ray flares remains a mystery. However, observations, especially with Chandra and Hubble, together with extensive computer simulations (figure 67), present the following general picture of the origin of the Crab Nebula: the core of a massive star collapsed and produced, as part of the supernova explosion process, a highly magnetized, rapidly rotating neutron star, or pulsar, which in turn produced a pulsar wind that flows away from the pulsar at near the speed of light, produces an extensive bubble of high-energy particles and magnetic fields, and is light-years in diameter.

The basic ingredients seem standard enough, almost off the shelf: a neutron star that is rapidly rotating and has a strong magnetic field. All that is needed to make this is a massive star, maybe one that is rotating a little faster than normal and has a magnetic field a little stronger than normal, but that should not be hard to come by in a galaxy containing a few billion massive stars.

One question comes immediately to mind: If it is relatively easy to make a pulsar wind nebula, then why weren't more of them known? The answer is that a telescope like Chandra was needed. Thanks to Chandra, it is now known that the Crab Nebula is just the most prominent example of a class of objects called pulsar wind nebulas, in which a rapidly rotating neutron star has created a cloud of high-energy particles that extends over several light-years. Chandra has found more than fifty such objects, and they appear to fall into two basic groups: those with a ring-jet geometry, such as the Crab Nebula, and another group with a swept-back, tadpole-like structure (figure 68).

The tadpole structure observed in many pulsar wind nebulas is produced by the rapid motion of the pulsar through the surrounding gas. The head of the tadpole is produced by a bow wave, similar to the wave produced by a duck paddling across a pond. Bow waves are a common feature of an object moving through a medium, be it a duck or a boat in a pond, airplanes moving through the sky, a comet moving through the interplanetary medium, or pulsars moving through interstellar gas. An object moving through any of these environments creates a series of pressure or sound waves. If the object moves fast enough, these waves merge into a three-dimensional bow wave that is called a bow shock. A familiar example of a bow shock in Earth's atmosphere is the sonic boom produced by the motion of a supersonic jet. As the airplane flies, it pushes on the air in front of it and creates sound waves. If the plane moves faster than the speed of sound, it creates a bow shock. People on the ground can hear the combination of these sound waves as the bow shock passes them in the form of a sonic boom.

Chandra's image of the pulsar wind nebula cloud known as the Mouse shows a stubby bright cloud of high-energy particles, about four light-years in length, swept back by the pulsar's interaction with interstellar gas (figure 69). The intense source at the head of the X-ray cloud is the pulsar, estimated to be moving through space at about 1.3 million miles per hour. It is as if the Crab Nebula were put in a cosmic wind tunnel and exposed to a supersonic wind. A cone-shaped cloud of radio-wave-emitting particles envelopes the X-ray column.

The Mouse, a.k.a. G359.23-0.82, was discovered in 1987 by radio astronomers using the Jansky Very Large Array. It gets its name from its appearance in radio images that show a compact snout, a bulbous body, and a long, narrow tail that extends for about fifty-five light-years. The origin of the pulsar's high velocity is unknown. It seems to be the rule rather than the exception that a pulsar

B_p/B_ϕ

4

1

0.25

10^{18} cm

67 | MAGNETIC FIELD STRUCTURE

The magnetic field structure in the Crab Nebula pulsar, shown in a high-resolution computer simulation. In what looks like a cosmic bad hair day, the magnetic field lines in the pulsar are colored according to their orientation. Sections with a dominating azimuthal, or equatorial, component ($B\phi$) are blue, and those with a dominating poloidal component (Bp) are red. Magnetic field components with intermediate orientations are green (see the color scale to the right). The length scale is on the left (one light-year is approximately 10^{18} cm). The surface of the termination shock is shown as the magenta area at center.

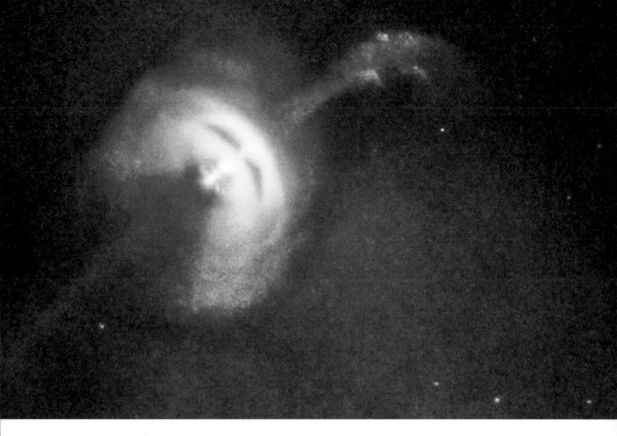

68 | VELA PULSAR WIND NEBULA

A Chandra image of the Vela pulsar wind nebula, about a thousand light-years from Earth, showing jets, arcs, and a swept-back look that owes to the pulsar's rapid motion through the supernova remnant in which it was created. The pulsar, or rotating neutron star, is the bright dot near the center of the lower of the two arcs created by the pulsar wind. The inner arc, likely a complete ring around the pulsar, is the location of a shock wave caused by the pulsar wind interacting with the outer nebula. The outer arc could be a feature similar to the Crab Nebula's larger ring. Jets produced by the pulsar flow out perpendicular to the arcs. The jets point in the same direction as the motion of the pulsar, which is about twelve miles in diameter and makes more than eleven complete rotations every second, faster than a helicopter rotor moves. As the pulsar whips around, it spews a jet of charged particles that race out along the pulsar's rotation axis at about 70 percent of the speed of light.

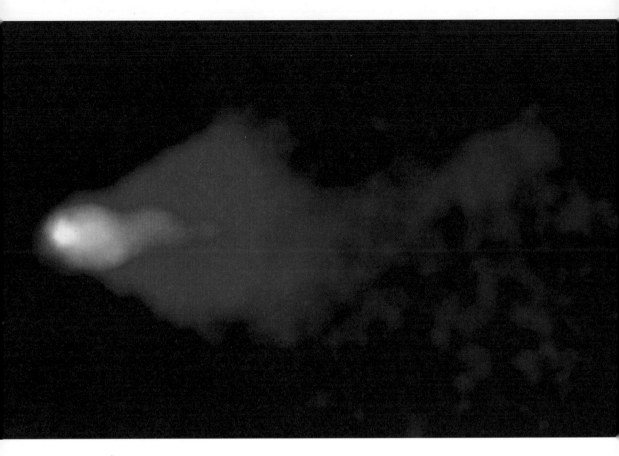

69 | THE MOUSE

The Mouse, a.k.a. pulsar wind nebula G359.23-0.82, about sixteen thousand light-years from Earth. Named for its appearance in radio images, it has a compact "snout," a rounded body, and a long, skinny tail trailing out about fifty-five light-years. The composite Chandra (pink) and radio (blue) images show a close-up of the Mouse's head, where a shock wave has formed as the young pulsar hurtles through interstellar space at about 1.3 million miles an hour.

receives a substantial kick in a supernova explosion, propelling it through space at a high velocity. One plausible explanation is that these pulsar kicks are related to the explosive circumstances involved in the birth of the pulsar. For example, an off-center supernova explosion could give the pulsar a kick. Chandra images of other pulsar wind nebulas illustrate the diversity of intriguing shapes. These shapes are all created by a rapidly spinning neutron star that is spewing out winds and jets of high-energy electrons as it travels at supersonic speeds through the interstellar medium and, in some cases, through the expanding remains of its parent star.

If that engenders pleasant thoughts of a butterfly emerging from a cocoon, consider another, less pleasant image: that of a black widow devouring its mate. The Black Widow pulsar is doing just that.

Known officially as pulsar B1957+20, the Black Widow has a pulsar wind nebula and is moving rapidly through interstellar space (figure 70). This motion creates a bow shock and a swept-back tail. In this respect it is similar to many of the other head-tail, tadpole-shaped pulsar wind nebulas discovered by Chandra. These are relatively young—thousands of years old—rapidly rotating neutron stars that have created an energetic wind and are moving rapidly through the galaxy.

The similarity ends there. The Black Widow is roughly a billion years old, and it could be older. It is a member of a class of neutron stars called millisecond pulsars, because they are rotating extremely rapidly, making an entire revolution in a few milliseconds. Think about that. An entire star about the size of Manhattan rotates around several hundred times a second, faster than the flicker of a light bulb.

How did this happen? Don't pulsars slow down because of the energy lost by radiation? It seems like a violation of the laws of physics—a neutron star keeps getting older and producing energy, yet it speeds up, thereby gaining energy. Are millisecond pulsars perpetual motion machines? Unfortunately, they are not. The key to this paradox, as it is to any system that appears to be a perpetual motion machine, is that there is energy being supplied to the system from outside. All millisecond pulsars are members of binary star systems, and the companion star supplies the energy. The steady push of infalling gas from the companion onto the neutron star spins it up in much the same way that pushing on a merry-go-round makes it rotate faster.

As the rotation speeds up, so does the pulsar dynamo effect of rapid rotation combined with a magnetic field, and in time a pulsar is born, or reborn. The rejuvenated pulsars become a strong source of X-ray and gamma ray radiation.

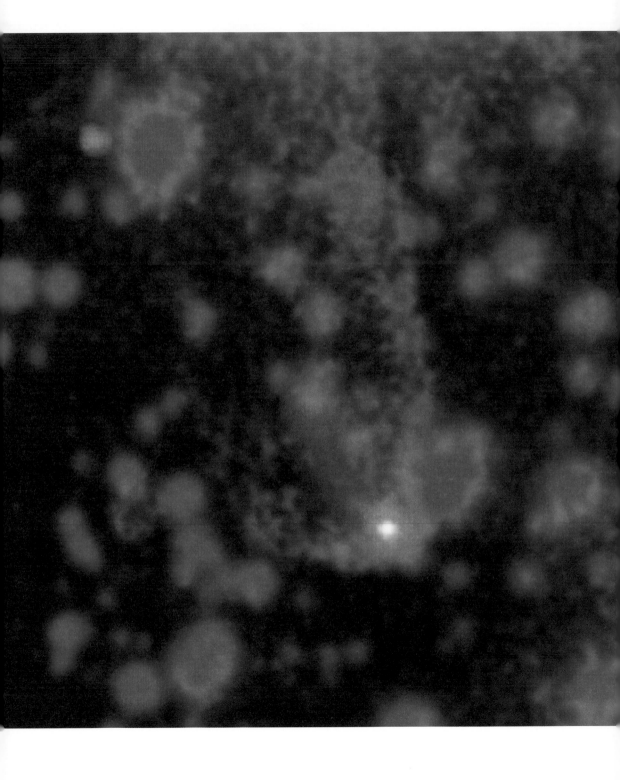

Some of the pulsar wind and high-energy radiation hits the companion star. The atmosphere of the companion star is heated and escapes from the star in a wind. This is not a minor effect. The companion star loses most of its mass and shrinks down to become a planet-sized brown dwarf star, at which point it can no longer provide "food" to the Black Widow. At that point, the Black Widow's rotation slows down and the pulsar emission declines, ending the feeding frenzy. But by this time, the companion has been devoured, having lost 90 percent or more of its mass.

70 | BLACK WIDOW

The Black Widow, or pulsar B1957+20. The pulsar is creating a bow shock wave visible at optical wavelengths (green) as it moves through the galaxy at a speed of almost half a million miles per hour. The pressure behind the bow shock creates a second shock wave that sweeps back the cloud of high-energy X-ray-emitting particles (red and white) produced by the pulsar to form the cocoon.

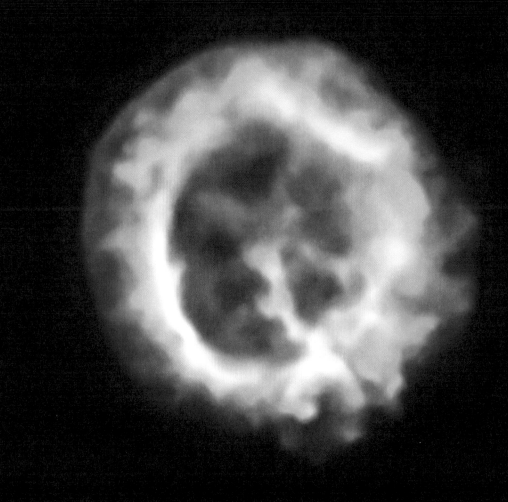

71 | E0102–72 SUPERNOVA REMNANT

A Chandra X-ray image of the remnant of a massive star that exploded in the Small Magellanic Cloud galaxy (ca. 190,000 light-years from Earth). The outer blue ring traces a shock wave moving at about 14 million miles per hour into the surrounding gas. The inner red-orange ring is expanding ejecta from the explosion that is being heated by a shock wave traveling back into the ejecta. The measured temperature of the shock-heated gas behind the forward shock is lower than expected, indicating that a significant fraction of the shock-wave energy has gone into accelerating particles to extremely high energies.

19

The results of my observation are best
explained by the assumption that a radiation
of very great penetrating power enters our
atmosphere from above.

Victor Hess, Nobel Lecture, 1936

A THIN COSMIC RAIN
PARTICLES FROM OUTER SPACE

Wilhelm Roentgen's discovery of X-rays in 1895 was a turning point in the history of physics. This was rapidly followed by Henri Becquerel's discovery of radioactive radiation, the discovery of radium by the Curies, and the quantum revolution in atomic physics. During this period, in their quest to understand the structure of the atom, scientists were bombarding all types of material with high-energy radiation, also called ionizing radiation because it would tear an electron off an atom, leaving behind a charged atom, or ion.

Scientists also searched for radioactive substances in the crust of Earth, in the ocean, and in the atmosphere. Radioactivity was found everywhere: in caves, in deep lakes, on high mountaintops. The most surprising discovery was that it was impossible to eliminate the influence of all radiation. No matter how thick the lead plates were that encased the detector, there was still a detectable signal. These rays were more energetic than any type of radiation previously observed and could penetrate lead plates a meter thick. Hess set out to discover the origin of this mysterious, energetic, natural radioactivity. He made significant improvements in the sensitivity and precision of his detectors, and then personally took the equipment aloft in a balloon. During 1911 and 1912, Hess systematically measured the radiation at altitudes above seventeen thousand feet. The daring flights were made both by day and during the night. He found that there was no significant difference between night and day, in or out of a solar eclipse. He did determine that the radiation detected at the highest altitudes he could reach was about twice that at sea level. The clear implication was that the radiation is coming from outer space. Robert Millikan, who later did research on the phenomenon, coined the term *cosmic rays* to describe it. In 1936, Victor Hess was awarded the Nobel Prize in physics for the discovery of cosmic rays.

In the decades that followed, evidence accumulated that cosmic rays are deflected by Earth's magnetic field, so they must be charged particles and not high-energy photons such as X-rays or gamma rays, which have no charge. So, in the sense of X-rays and gamma rays, which are high-energy forms of electromagnetic radiation, cosmic rays are not rays—they are particles, like alpha rays (helium nuclei) and beta rays (electrons). But the name stuck, and they are still called cosmic rays. During the years from 1930 to 1945, a wide variety of investigations confirmed that the primary cosmic rays are positively charged moving near the speed of light. They are mostly protons, with a mixture of approximately 10 percent helium nuclei and 1 percent heavier nuclei of elements such as carbon and iron. This composition is very similar to the abundances of elements found in the Sun and in interstellar space.

Spaceship Earth makes a journey of 500 million miles around the Sun each year. Throughout this journey, a thin rain of cosmic rays produces a background of ionizing radiation on Earth. During the Apollo flights, astronauts reported seeing odd flashes of light, visible even with their eyes closed. We have since learned that the cause was cosmic rays.

Cosmic rays are a definite hazard for astronauts, as they are for passengers and crew on high-altitude airplane flights, especially during solar flares, when the radiation doses from cosmic rays can reach dangerous levels. Even on Earth, our bodies are continually bombarded by cosmic rays, which damage or disrupt our molecules, and are thereby potentially harmful to our health. Another bit of bad news is that cosmic rays can, and do, cause glitches in the operation of our computers.

On the positive side, all this cosmic disruption can be put to good use to help answer questions and solve problems. As cosmic rays tear through Earth's upper atmosphere, they initiate nuclear reactions that change the nuclei of certain atoms. Some of these mutated nuclei, in the form of carbon-14 nuclei—so called because the nucleus contains six protons and eight neutrons, for a total of fourteen nucleons—can be identified and used as clocks to date the charred embers of ancient campfires, as well as bones and cloth.

Recently, with the development of powerful new mass spectrometers (particle accelerators that whirl nuclei around at dizzying speeds and use their orbits to determine their masses), scientists are using nuclei produced by cosmic rays—called cosmogenic nuclides—to study the history of Earth itself.

The principle is similar to that used for carbon-14. Cosmic rays produce cosmogenic nuclides of beryllium-10 and aluminum-26 in quartz crystals of rock and sand while they are exposed to the cosmic weather. The longer a quartz crystal is exposed, the more cosmogenic nuclides it accumulates. Isotopes formed by collisions with cosmic rays in exposed rock and sand crystals are washed down into streams, where geologists can collect the sediment. This technique has been used by Darryl Granger of Purdue University to show that the ice sheet that formed Mammoth Cave in Kentucky arrived about 1.5 million years ago, about 700,000 years earlier than previously thought. At the other end of the scale, a group led by Friedhelm von Blanckenburg in Switzerland found that the onset of cultivation in the rain forest of Sri Lanka increased erosion as much as a hundredfold.

Over the long haul, one might think that mountains wear down at a fairly steady rate. Not so, discovered Jim Kirchner of the University of California, Berkeley. His study of Idaho's mountains revealed that erosion rates over the past five thousand to twenty-seven thousand years were seventeen times higher than modern-day rates. Kirchner concluded that rare catastrophic events, such as wildfires followed by floods, were responsible for most of the erosion. This information is useful for engineers in estimating the time it will take reservoirs to fill with debris.

That cosmic-ray dating techniques should demonstrate the importance of catastrophes in shaping Earth is fitting in a way, because most of the cosmic rays that rain down on Earth are probably due to cosmic catastrophes: supernovas.

Supernovas produce brilliant flashes of light and awesome shock waves that rip through interstellar space at speeds of millions of miles per hour. Part of the energy of a supernova shock wave goes into the production of hot gas that can be observed with an X-ray telescope (figure 72). Another portion goes into the acceleration of electrons, whose radiation is observed with radio, optical, and X-ray telescopes.

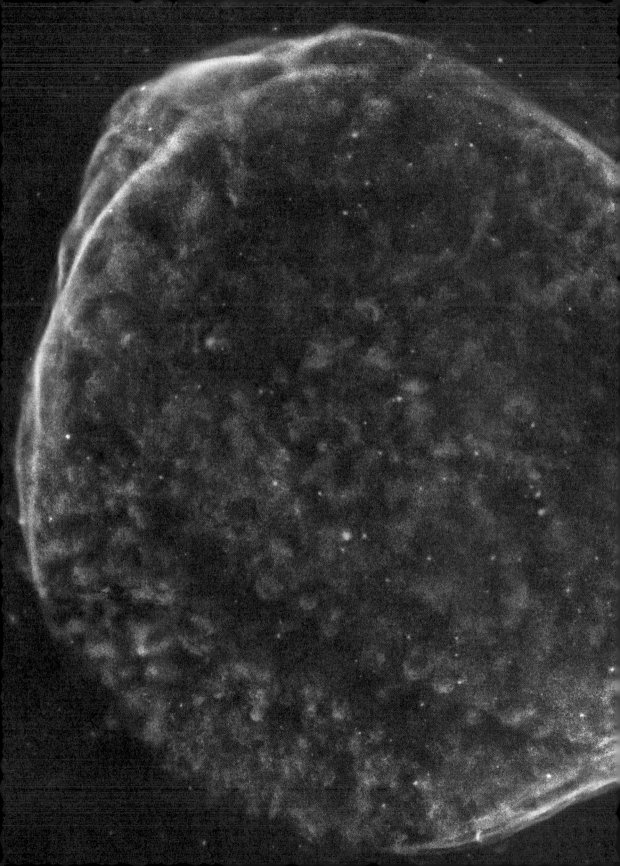

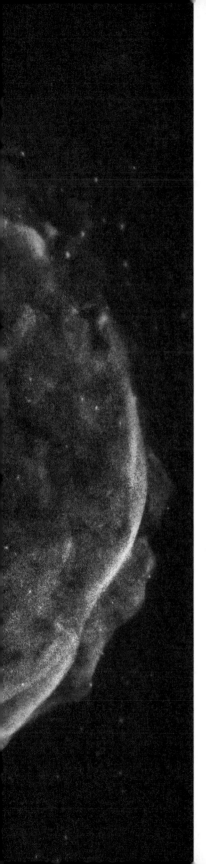

72 | SN 1006 SUPERNOVA REMNANT

The remnant of the Type Ia supernova, seven thousand light-years from Earth, that astronomers in China, Japan, Europe, and the Arab world observed in 1006 CE. Low-, medium-, and high-energy X-rays detected by Chandra in this composite of ten views are red, green, and blue, respectively. The light-blue rim traces the forward shock. The proximity of the forward shock wave to the ejecta indicates that a significant amount of the wave's energy has gone into accelerating cosmic rays. A similar effect is observed in the Tycho (figures 52 and 73), E0102-72 (figure 71), and Cassiopeia A supernova remnants.

Not long after it had been demonstrated that cosmic rays are charged particles, Walter Baade and Fritz Zwicky suggested that supernovas were the source of cosmic rays. Later, in a classic 1964 monograph on cosmic rays, Russian physicists Vitaly Ginzburg and Sergei Syrovatskii put this suggestion on much stronger footing. However, until recently, there was no direct evidence as to how protons and heavier nuclei are accelerated to speeds approaching the speed of light. Because cosmic rays are composed of charged particles, like protons and electrons, their direction of motion changes when they encounter magnetic fields throughout the galaxy. The origin of individual cosmic rays detected on Earth therefore cannot be determined.

Chandra has helped to illuminate how supernovas produce cosmic rays. X-ray images of the Tycho supernova remnant show that the shock waves produced by the supernova behave in an unexpected way. The supernova debris is observed to expand at a speed of about 6 million miles per hour. This rapid expansion has created two X-ray-emitting shock waves: the forward shock wave, moving outward into the interstellar gas, and the reverse shock wave, moving inward into the stellar debris. These shock waves, analogous to the sonic boom produced by supersonic motion of an airplane, produce sudden, large changes in pressure and temperature behind the wave. According to the standard theory—before the Chandra observations—the forward shock wave should be about two light-years ahead of the stellar debris (half the distance from our Sun to the nearest star). What Chandra found instead is that the stellar debris has kept pace with the forward shock and is only about half a light-year behind. Similar effects have been observed for other young supernova remnants.

The most likely explanation is that a large fraction of the energy of the forward shock wave is going into the acceleration of cosmic rays, as seen in figure 72. Previous observations with radio and X-ray telescopes had established that the shock wave in Tycho's remnant was accelerating electrons to high energies.

However, since high-speed atomic nuclei produce very weak radio and X-ray emission, it was not known whether the shock wave was accelerating atomic nuclei as well. The Chandra observations provide the strongest evidence yet that nuclei are indeed accelerated, and that the energy contained in high-speed nuclei is about a hundred times that in the electrons. The detection of gamma rays from the Tycho supernova remnant supports this conclusion. The intensity of the gamma rays indicates that they are likely due to collisions between high-energy protons with interstellar atoms. These results, and similar ones from other supernova remnants, provide strong support for the idea that supernovas are the major source of cosmic rays that impact Earth. Extremely high-energy cosmic rays likely come from quasars produced by matter swirling into supermassive black holes, or possibly from shock waves created by intergalactic matter falling into clusters of galaxies.

Based on the successes from the observations of the Tycho supernova remnant, scientists took an even longer look at it with Chandra. The new image of Tycho's supernova remnant revealed a peculiar pattern of stripes near the outer edge of the remnant (figure 73). Assuming that the spacing between the X-ray stripes corresponds to the radius of the spiraling motion of the highest-energy protons in the supernova remnant, the spacing corresponds to energies more than thirty times higher than reached in the Large Hadron Collider.

The discovery of the X-ray stripes provides support for a general theory according to which magnetic fields can be dramatically amplified and tangled in supernova shock waves. The X-ray stripes are thought to be regions where the magnetic fields are more tangled than surrounding areas. Electrons become trapped in these regions and emit X-rays as they spiral around the magnetic field lines. However, the theory did not predict the regular and almost periodic pattern of the X-ray stripes, and clearly this is an important clue for further theoretical study.

In the meantime, what Chandra has clearly established is that supernovas in our galaxy produce many of the cosmic rays that affect our lives and inform us about life on Earth. The cosmic environment that we inhabit profoundly affects us. Maybe not in the way the daily astrology columns portend, but in real, practical ways. Some of these effects may be the source of "Earthly" problems, and some of them can be used in the search for solutions to problems.

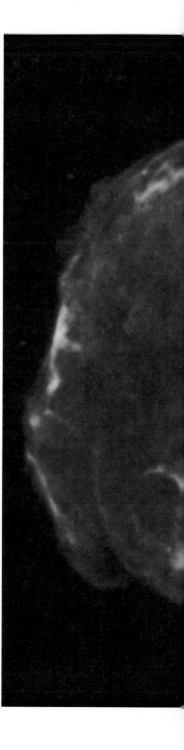

73 | TYCHO SUPERNOVA REMNANT

X-ray stripes in Tycho SNR, the first ever spotted in a supernova remnant. This image depicts only the higher-energy X-rays Chandra detected from the Tycho SNR (see also figure 52). These X-rays, produced by extremely energetic electrons, show the supernova's forward shock wave. The two insets show (A) the region containing the brightest stripes, and (B) a region with fainter stripes. The stripes are regions where the magnetic fields are tangled and particle motion is much more turbulent than in surrounding areas. Electrons, trapped in these regions, emit X-rays as they spiral around the magnetic field lines. The spacing between the stripes corresponds to the radius of the spiraling motion of a proton with an energy more than thirty times higher than that of the Large Hadron Collider in Switzerland. Very energetic protons do not radiate X-rays efficiently, and Chandra cannot detect them, but they have been observed with gamma ray telescopes. Acceleration of protons in supernova shock waves may play an important role in creating the cosmic rays detected on Earth.

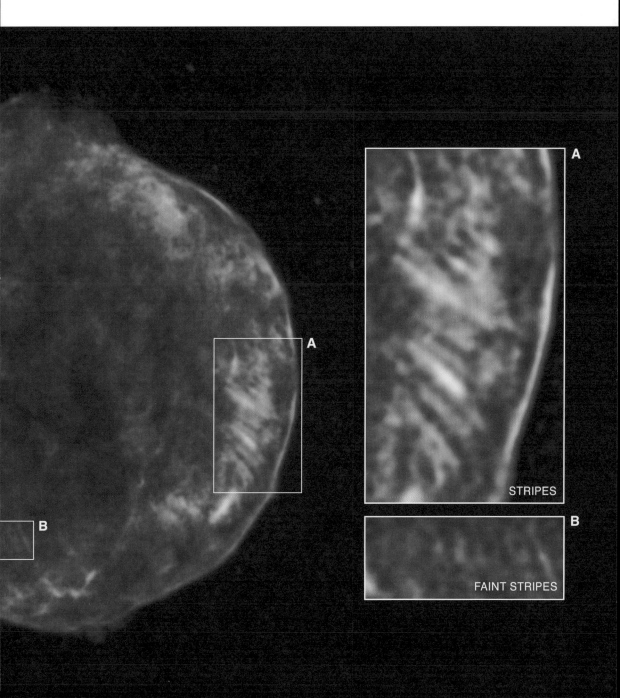

A

B

A

STRIPES

B

FAINT STRIPES

74 | ANTENNAE GALAXY SYSTEM

An optical image. The long, antenna-like arms of stars and gas seen in wide-angle views such as this one give the system its name. These features were produced by gravitational forces as two galaxies collided.

20

What has been the history of matter? . . .

This history is hidden in the abundance

distribution of the elements.

E. Margaret Burbidge, Geoffrey R. Burbidge, William Fowler, and Fred Hoyle,
"Synthesis of the Elements in Stars," 1957

ELEMENTAL CHANGE

Chemistry, the study of the intricate dances and pairings of low-energy electrons to form molecules of the various compounds that make up the world we live in, may seem far removed from the thermonuclear heat in the interiors of stars and the awesome power of supernovas. Yet there is a fundamental connection between them.

An atom of ordinary or baryonic matter, as it is called to distinguish it from the still mysterious dark matter, is composed of a cloud of negatively charged electrons orbiting a nucleus composed of protons and neutrons. Protons have a positive charge, and neutrons, as the name suggests, are electrically neutral. Although the nucleus of an atom has a diameter of only one one hundred thousandth that of the atom's electron cloud, the nucleus contains more than 99.9 percent of the mass of an atom. The number of protons in the nucleus defines chemical elements. All hydrogen atoms all have one proton in the nucleus, all helium nuclei have two protons, all carbon nuclei have six protons, and so on. Elements can have different isotopes; that is, the number of neutrons in the nucleus can differ—for example, a carbon-12 nucleus has six protons and six neutrons, whereas a carbon-14 nucleus has six protons and eight neutrons.

The formation of the elements began about 14 billion years ago, in the early minutes of the Big Bang. About one minute after the Big Bang, almost all the baryons in the universe were in the form of protons, the nuclei of the simplest element, hydrogen, with a mixture of neutrons. By the time the universe was twenty minutes old, the baryonic matter was a weighted mixture of 75 percent hydrogen nuclei, 25 percent helium nuclei composed mostly of two protons and two neutrons, and free-ranging electrons. A few hundred thousand years later, the negatively charged electrons and positively charged nuclei would form neutral atoms. In essence, the history of the formation of the elements can be divided into two main phases: one that ended after the first twenty minutes, and another that has been ongoing since the formation of the first stars 13-plus billion years ago.

After that initial one-third of an hour, the expanding universe cooled below the point where nuclear fusion could operate. No further evolution of matter could occur again until hot, dense places were created again. This did not happen until a few million years later, when the first stars formed. Then the buildup of elements heavier than helium could begin deep in the interiors of stars. There, the gravitational crush of overlying material compressed the gas and raised temperatures to millions of degrees. Nuclear fusion reactions turned on, fusing helium nuclei into carbon, carbon into oxygen, and so on. Winds of gas escaping from stars distribute some of this processed matter into space in a relatively gentle manner, and supernovas do it violently.

This process was first described in detail in a monumental 104-page paper published in 1957 by E. Margaret Burbidge, Geoffrey R. Burbidge, Fred Hoyle, and William Fowler. "Synthesis of the Elements in Stars" described how stars synthesize nearly all the chemical elements heavier than hydrogen and helium, all the way up to uranium. This work is generally agreed to be one of the most important papers ever in astrophysics. It laid the foundation for a new kind of synthesis of astronomical observations with nuclear and particle physics and paved the way for much of modern astrophysics.

In the introduction to the paper (which became known as B^2FH, for "B-squared FH" after the initials of their last names), the authors state that their goal is to determine the "history of matter." This history is hidden in the abundance distribution of the elements, that is, the relative proportion of each element in the universe.

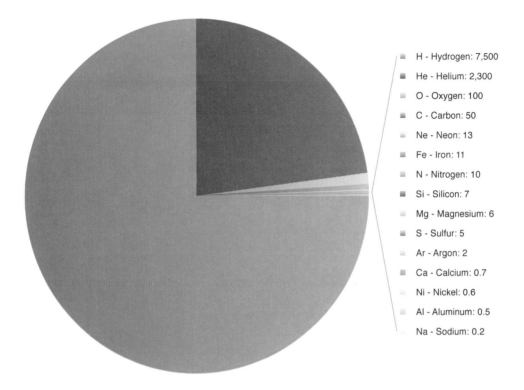

Legend:

- H - Hydrogen: 7,500
- He - Helium: 2,300
- O - Oxygen: 100
- C - Carbon: 50
- Ne - Neon: 13
- Fe - Iron: 11
- N - Nitrogen: 10
- Si - Silicon: 7
- Mg - Magnesium: 6
- S - Sulfur: 5
- Ar - Argon: 2
- Ca - Calcium: 0.7
- Ni - Nickel: 0.6
- Al - Aluminum: 0.5
- Na - Sodium: 0.2

75 | COSMIC ABUNDANCE OF ELEMENTS

A pie chart illustrating the average abundance by mass of the various elements in the universe (called the cosmic abundances) in parts per ten thousand. Note the large abundance of hydrogen and helium and the small abundances of all the rest of the elements (about 2 percent of the total). This, as well as evidence from other observations, indicates that the universe was once composed of just one element, hydrogen, and that the nuclear fusion reactions that produce heavier elements have proceeded very slowly over the universe's 13.7-billion-year history. Earth does not have such large abundances of hydrogen and helium because these lighter elements boiled off as the planet formed.

A pie chart of the abundances of the elements (figure 75) shows that the simplest elements, hydrogen and helium, are far and away the most abundant, with oxygen and carbon coming in a distant third and fourth. This preponderance of hydrogen and helium is strong evidence that the elements in the universe have been created by a process in which the heavier elements are built up from the lighter ones, starting with hydrogen. This process is called nucleosynthesis.

As stars evolve, depending on their mass, they go through a sequence of stages in which nuclear fusion reactions build up heavier elements and, in the process, supply energy to support the star against the crushing force of gravity. Ultimately, processed matter containing heavy elements—any element heavier than helium is considered a heavy element by astrophysicists—is expelled into the interstellar medium either by stellar winds or by supernova explosions. One of the principal scientific accomplishments of Chandra has been to help unravel exactly how the enrichment by stellar winds and supernovas works on a galactic and intergalactic scale.

Chandra images and spectra of individual supernova remnants make it possible to track the speed at which clouds of gas rich in elements such as oxygen, silicon, sulfur, calcium, and iron were ejected in the explosions. In some cases, it is possible to construct three-dimensional images, which will be key to developing accurate models for the explosion process.

On a galactic scale, observations indicate that stars are forming extremely rapidly in certain peculiar galaxies, called starburst galaxies. Starbursts are apparently triggered by collisions with other galaxies. These collisions initiate the collapse of interstellar clouds.

In the Antennae galaxy system (figures 74 and 76), a collision of two galaxies has created several bursts of star formation. When galaxies collide, the stars in the galaxies do not—their target area is too small—but large, diffuse clouds of gas do, and they did collide frequently in past events. These collisions sent shock

76 | ANTENNAE GALAXY SYSTEM

A Chandra image. The collision of galaxies that produced this system created a burst of star formation that eventually led to enhanced supernova activity. This image shows clouds of hot interstellar gas enriched with oxygen, iron, silicon, and other heavy elements ejected from many supernovas in the system. The bright, pointlike sources are caused by black holes and neutron stars, also the product of supernovas.

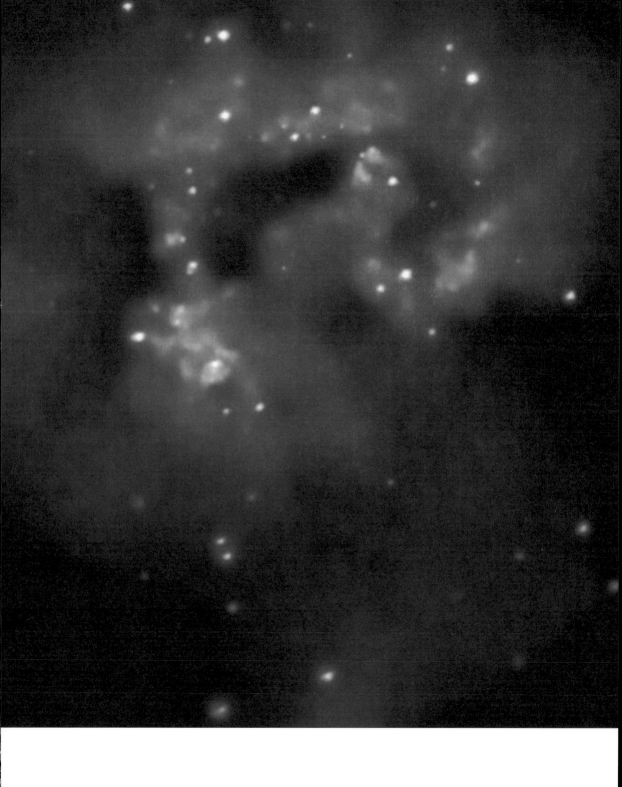

waves rumbling through the clouds. Tens to hundreds of millions of stars formed in a few million years, a "lost weekend" or so in the life of a galaxy. Millions of years later, this "baby boom" leads to a "death boom" in the form of many supernovas. These heat the interstellar gas and enrich it with heavy elements.

Another, unexpected agent for distributing heavy elements throughout a galaxy is a supermassive black hole in the center of the galaxy. Violent winds and jets produced by rapidly spinning black holes can drive material enriched with heavy elements into the outer reaches of the galaxy and beyond.

Gas can also be stripped from a galaxy if the galaxy falls into a galaxy cluster. The motion of the galaxy through the hot intergalactic gas produces a ram pressure that acts like a wind, in much the same way that a motorcyclist traveling at a high speed feels a strong headwind (figure 77). This so-called ram pressure wind blows the cool gas out of the galaxy into the intergalactic medium and enriches it with the heavy elements produced by stars in the infalling galaxy.

On a still larger scale, evidence for oxygen has been detected in intergalactic filaments of gas millions of light-years in length. This oxygen was likely produced more than 10 billion years ago, in some of the first supernovas to occur in the evolution of the universe. In some starburst galaxies, the rate of supernovas gets so high that the combined effects of many supernova shock waves drives a galactic-scale wind that blows the gas out of the galaxy. A prime example is the galaxy M82, shown in figure 78.

Galactic winds such as the one in M82 are rare today, but they were probably common billions of years ago, when galaxies were very young. Running the expansion of the universe backward to that time implies that galaxies were closer together then. Collisions between galaxies were much more common than in the present epoch. These collisions produced more starbursts, and millions of years later more supernovas, which overheated the interstellar gas to the point that it expanded out of the galaxy into intergalactic space. A Chandra report on the detection of oxygen ions in the Sculptor Wall, a vast collection of gas and galaxies that

77 | RAM PRESSURE

A composite Chandra X-ray (blue) and Hubble optical (cyan, orange, and white) image of the spiral galaxy ESO 137-001, more than 200 million light-years away. This image shows streaks of gas that ram pressure stripped off the galaxy. ESO 137-001 is moving toward the upper left of this image, between other galaxies in the Norma cluster.

stretches across millions of light-years, may be evidence that this process occurs on a vast scale. As the enrichment of the interstellar and intergalactic gas in heavy elements has proceeded over vast stretches of space and eons of time, the chemistry of the cosmos has become more complex. Subsequent generations of stars have formed from interstellar gas with more heavy elements. Our Sun, solar system, and indeed the existence of life on Earth are direct results of this long chain of stellar birth, death, and rebirth. In this way, the evolution of matter, stars and galaxies are all inextricably tied together, and so too are astronomy and chemistry.

The message is both simple and profound. As time passes, things change. The elements, which are the bedrock of our existence, change into other elements. Impermanence rules. Stars are born, exist for a while—sometimes

78 | M82 GALAXY

A starburst galaxy roughly 12 million light-years from Earth. While optical light from stars (yellow and green, from Hubble) reveals the disk of an evidently normal galaxy of modest size, a second Hubble observation, which images ten-thousand-degree Celsius hydrogen gas (orange), paints quite a different picture: matter is blasting out of the galaxy. A Spitzer infrared image (red) shows cool gas and dust also ejected outward. Finally, Chandra's X-ray image (blue) shows gas heated to millions of degrees by this outflow. The eruption originates in M82's central regions, where stars are forming at a rapid clip— sometimes ten times faster than in the Milky Way. Ejection of matter from galaxies during peaks in star formation might be a major way that elements such as carbon and oxygen disperse through the universe. M82's surge of star formation was probably spurred circa 100 million years ago when M81, another big nearby galaxy, approached M82.

for a very long while—but then they explode, implode, or just slowly cool into obscurity.

Of course, on the human scale, we knew this already. It is the ecstasy and agony of our existence. We are born, we live, we love, friends and loved ones die, and so do we.

And yet . . . A few years ago, I attended a memorial service for Geoffrey Burbidge, who was my Ph.D. thesis advisor a long time ago. A colleague whom I had not seen for years remarked, "People don't really change that much, do they? They grow old, but the fundamentals are still there. You can tell they're still the same person."

Geoff certainly didn't change. To the end, he steadfastly stuck to contrarian points of view on the Big Bang and related topics, which pushed him out of the mainstream of physics, where he had once been such a commanding figure. On a more personal level, he remained a gruff, intimidating presence—I became his graduate student because the other students avoided him, and I, being green, fresh from Oklahoma, and clueless, didn't know any better. As it turned out, it was a stroke of luck for me. He became a mentor who guided me into X-ray astronomy and helped me get my first job in astrophysics. Then he became a friend who, when I was in the depths of despair and helpless to prevent a catastrophic change, was one of the first to call me and offer his friendship, support, and consolation, even though he himself was ill.

Simple acts of kindness and consideration. They may not seem like much, but they can offer something of unchanging value as we struggle to cope with a changing universe.

79 | W44 SUPERNOVA REMNANT

A composite of Chandra X-ray (blue) and Spitzer (red and green) images. Known as G34.7-0.4, W44 is a supernova remnant evolving inside an interstellar molecular cloud. The Chandra image reveals that the remnant's interior is rich in heavy elements synthesized by nuclear reactions in the star over the course of its lifetime. These elements are being dispersed into the galaxy by the supernova, and some eventually will be assimilated into new stars that form from the cloud. W44 is also the source of gamma ray emission produced by high-energy protons produced in the supernova shock wave interacting with the surrounding gas.

80 | ORION NEBULA CLUSTER

At a distance of about 1,400 light-years, this is one of the closest star-formation regions to Earth. Chandra observed Orion almost continuously for thirteen days, a long look that enabled scientists to study the X-ray behavior of young Sun-like stars with ages between 1 and 10 million years. They discovered that such stars produce violent X-ray outbursts, or flares, that are much more frequent and energetic than anything seen today from our 4.6 billion-year-old Sun. This suggests that the Sun may have been much more active in its youth. Orange, yellow, and blue represent low-, medium-, and high-energy X-rays detected by Chandra.

21

What was before

Is left behind; what never was is now;

And every passing moment is renewed.

Ovid, *Metamorphoses*, 8 CE

COSMIC RENEWAL

E xcept for the studies of the cosmic microwave background radiation, virtually all research in cosmology and extragalactic astrophysics depends on knowledge of how stars form and evolve. Yet, despite considerable progress in characterizing and understanding star formation in isolation and in small molecular clouds, knowledge of star formation is limited, with fundamental questions still unanswered: What is the relative proportion of low-, medium-, and high-mass stars that form when a huge cloud of dust and gas collapses and fragments into stars? What exactly triggers the formation of stars? How many stars have disks that will be the birthplace of planets?

One of the first steps toward answering these questions is to observe interstellar clouds at various stages in the star formation process. To use an old but good analogy, astronomers have learned to piece together the essence of the billion-year drama of stellar evolution. They have done this in much the same way that an observant biologist could understand the general outlines of the life cycle of trees in the forest by spending in a day in the forest, even though the trees may live more than a hundred years. The biologist could take note of the location and numbers of acorns or pinecones, sprouts, and trees at various stages of growth.

In the same way, astrophysicists can observe stars of different ages and masses and reconstruct the life cycles of stars (figure 81). Extensive catalogs of stars have been compiled, listing properties of the stars such as luminosity, color, and so on, and these data have been used to develop the theory of stellar evolution. It turns out that there are many more low-mass stars than massive ones. At the very lowest end of the mass scale are brown dwarfs, which have a mass of only a few percent of that of the Sun. Brown dwarfs cannot sustain nuclear reactions, so they never evolve, and they spend their lives in a limbo state between that of a planet and a star. The faintness of brown dwarfs makes them difficult to detect and to assess just how abundant they are in the galaxy.

The formation of a star begins with the collapse of a cloud of dust and gas. In the early stages, the energy released by gravitational collapse heats the interior of the star. This is the so-called protostar stage of star formation. After a few million years, the material in the center of the protostar becomes so dense and hot that the nuclear fusion of hydrogen nuclei into helium nuclei can occur. The outflow of energy released by these reactions provides the pressure necessary to halt the collapse. A star is born. Stars progress through their lives as balls of gas heated by thermonuclear reactions in their cores. If they have masses less than a few Suns, they will end their lives as white dwarf stars. The more massive stars explode as supernovas, leaving behind neutron stars or, in the case of the most massive stars, black holes. Indeed, the development of an understanding of how stars evolve is one of the great triumphs of twentieth-century astrophysics, if not all science.

Many details remain to be worked out, especially on the front end, when stars form. In what follows, I lump protostars and young stars together as young stars, since it is difficult to assess the precise boundary between the two stages. What is known is that stars do not form in isolation but in clusters, which can have thousands of members. Because the stars in a cluster were all born at roughly the same time and are at roughly the same distance from Earth, a star cluster provides an ideal laboratory for testing theories of how the behavior of a star relates to its mass. Most star clusters eventually disperse, so we see them when the stars are relatively young, with ages less than a few hundred million years.

One problem is that much of the star formation process takes place in the dark, inside dense clouds of gas and dust. The dust in particular absorbs optical

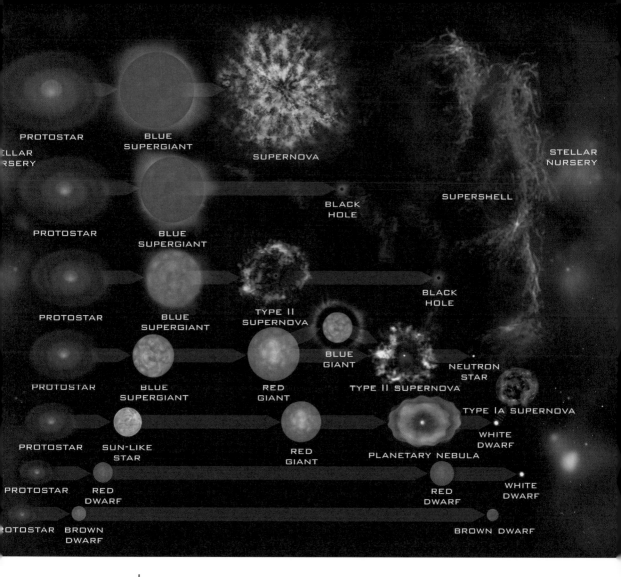

PROTOSTAR   BLUE SUPERGIANT   SUPERNOVA   STELLAR NURSERY

STELLAR NURSERY

BLACK HOLE   SUPERSHELL

PROTOSTAR   BLUE SUPERGIANT   BLACK HOLE

PROTOSTAR   BLUE SUPERGIANT   TYPE II SUPERNOVA   BLUE GIANT   NEUTRON STAR

PROTOSTAR   BLUE SUPERGIANT   RED GIANT   TYPE II SUPERNOVA   TYPE Ia SUPERNOVA

WHITE DWARF

PROTOSTAR   SUN-LIKE STAR   RED GIANT   PLANETARY NEBULA

PROTOSTAR   RED DWARF   RED DWARF   WHITE DWARF

PROTOSTAR   BROWN DWARF   BROWN DWARF

81 | EVOLUTION OF STARS

A summary of the evolution of stars, from birth through middle age, old age, and death. The lowest-mass stars are at the bottom and the highest-mass ones at the top. The types of evolution in the top two rows are theoretically possible but have yet to be confirmed by observations. Stars are formed in giant clouds of dust and gas and progress through their normal life as balls of gas heated by thermonuclear reactions in their cores. The rate of evolution and the ultimate fate of a star depend on its mass, with more massive stars evolving more rapidly than low-mass ones. Brown dwarfs, with a mass only a few percent that of the Sun, cannot sustain nuclear reactions and thus never evolve. Depending on its mass, a star ends its evolution as a white dwarf, neutron star, or black hole. The cycle begins anew as an expanding supershell from one or more supernovas triggers the formation of a new generation of stars.

light so efficiently that even the most powerful optical telescopes cannot see what is happening there. As a look at almost any of the beautiful images taken with optical telescopes shows, there are also a lot of stars in the field of view, most of which have nothing to do with the star-forming cloud. They constitute an attractive nuisance as far as understanding star formation is concerned.

Fortunately, help has arrived in the form of Chandra, with its ability to make sensitive, high-resolution X-ray images of star clusters and their environs. X-rays are largely unaffected by dust, and it helps that young stars are fairly strong X-ray emitters. The galactic field stars in the foreground are mostly older, weak X-ray emitters, so they are a minor problem. Chandra's spatial resolution makes it possible to distinguish individual stars, something other X-ray astronomy missions could not do. Previously unseen populations of stars, embedded in gas and dust and invisible to optical telescopes, have been discovered.

At a distance of about 1,400 light-years, the Orion Nebula Cluster is the closest large star-forming region to Earth (figure 80, page 224). Chandra images show about a thousand X-ray-emitting young stars in the Orion Nebula star cluster. The X-rays are produced in the hot, multimillion-degree upper atmospheres, or coronas, of these stars. Multiple Chandra observations of the Great Nebula in Carina, a star-forming region in the southern hemisphere, have yielded a catalog of more than fourteen thousand X-ray point sources (figure 82). More than twelve thousand of these are young stars, with ages between 1 million and 10 million years.

One significant advance made possible by these observations is the development of a method to estimate stellar ages in young star clusters, based on a combination of X-ray and infrared observations and theoretical models for the evolution of protostars and young stars. This method has been applied by a team of scientists from Penn State University to derive median ages for thousands of stars in seventeen massive star-forming regions. They found that star-forming regions are not uniformly populated with young stars of the same age. Rather, the stars form in clumps or subclusters. Large differences in stellar ages across the clusters indicate that star formation occurs in waves.

Detailed Chandra and infrared studies of the Flame Nebula (NGC 2024) and Orion Nebula Cluster show that the progression of star formation in a large interstellar cloud can be traced over spatial scales of several light-years and time scales of several million years. In the Flame Nebula (figure 83), the stars in the center have ages of about two hundred thousand years, whereas those on the outskirts are about 1.5 million years old. The data imply that some star clusters

CHANDRA'S COSMOS

form slowly, that the apparent age ranges are real, and that star clusters do not necessarily form from the inside out. For example, if a supernova happens to occur near the edge of a cloud, it can drive a shock wave, which triggers a wave of star formation. Sometimes the radiation from an extremely luminous star before it goes supernova is sufficient to drive a shock wave and initiate star formation in the vicinity.

A major finding of recent research, primarily with infrared telescopes, is that very young stars are usually surrounded by accretion disks of dust and gas. As with disks around black holes, these disks are a natural consequence of the collapse, with random gas motions averaging out to go with the flow of angular momentum, producing a flat disk like a pizza. Some of this matter will fall from the disk onto the young star. The accretion of gas may produce X-rays as the particles collide with the atmosphere of the star. As with accretion into black holes, the combination of rotation and magnetic fields may produce a jet of gas streaming away perpendicular to the accretion disk.

Circumstellar disks are the raw material from which planets will form, so they are of great interest in efforts to understand how planets are formed. An important question is the survival of circumstellar disks in the often harsh conditions of a star-forming cloud. A deep Chandra observation, combined with optical and infrared studies, of the approximately 3-million-year-old Trumpler 16 star cluster in the Carina Nebula investigated this question. It revealed that a large fraction of the circumstellar disks around low-mass stars have survived the intense radiation field of the cluster's massive, luminous stars. However, the disks around the massive stars do not fare so well, and the survival rate of these disks is several times lower than that of the low-mass stars. Chandra and infrared surveys found that the disk fraction for stars with masses greater than twice that of the Sun is about 20 percent, whereas the fraction for stars with masses comparable to that of the Sun is about 50 percent. Disks around massive stars evidently have a shorter dissipation timescale than those around lower-mass stars. Even for low-mass stars, radiation and the disruptive influence of the gravity of the young star take their toll. Comparative studies of many star clusters suggest that the fraction of stars with disks is about 80 to 90 percent in young clusters less than about a million years old. By 5 million years, the disk fraction drops to 20 percent, and after 10 million years, almost all disks have dissipated, having been blown away or accreted onto their parent star. This means that the formation of gas planets has to be completed within about 10 million years, before all the material that will form the planets is gone.

Now that the time scale for the destruction of circumstellar disks has been fairly well established, it is possible to flip the relationship and use the percentage of disks around stars as method for estimating the age of a young star. A young star is much more likely than an old one to have a disk. Chandra, optical, and infrared observations of the massive young cluster NGC 6611 and its parental cloud, the Eagle Nebula, indicate that waves of star formation have swept across this nebula. A giant shell of cool molecular hydrogen gas about six hundred light-years wide likely was created by a wave of supernova activity 6 million years ago, followed by another wave 2 to 3 million years ago and a third wave three hundred thousand years ago.

Strong X-ray emission from young stars indicates that a significant amount of matter is accreting from the circumstellar disk onto the star. This accretion, which is stronger for young stars less than a few million years old, is likely guided by a magnetic field connecting the disk and star. The magnetic field can become twisted and tangled in this process. In addition to producing jets, it can cause intense flaring on the star.

A long Chandra observation of the Orion Nebula star-forming region has shown that virtually all young stars exhibit powerful flares up to ten thousand times stronger than the flares on the present-day Sun. A strong feedback loop could be at work around these stars. Intense flares, occurring typically once a week, can heat the circumstellar disk and ionize it, creating more turbulence in the disk, which creates more accretion, which produces more flaring. It is not

82 | GREAT NEBULA IN CARINA

A relatively nearby star-forming region in the Sagittarius-Carina arm of the Milky Way, about 7,500 light-years from Earth. Chandra has detected more than fourteen thousand stars in this area, revealed a diffuse X-ray glow, and provided strong evidence that massive stars have already self-destructed as supernovas. The lower-energy X-rays in this image are red, medium-energy X-rays are green, and the highest-energy X-rays are blue. The diffuse X-ray emission shown here comes from a combination of supernova remnants and gas heated by winds from massive stars, some of which can be seen as the bright X-ray sources scattered across the image. These stars will explode as supernovas in a few million years or even sooner.

TRUMPLER 15

TRUMPLER 14

ETA CARINA

TRUMPLER 16

known what the ultimate effect will be for the formation of planets. On the one hand, the process could consume most of the gas before planets could form. On the other hand, the increased turbulence could encourage planets to form.

The X-ray study of exoplanets is rapidly developing in an effort to answer these and other questions. In the case of giant planets in the solar system, located several hundred million kilometers from the Sun, residual heat from formation plays an important role in their energy budgets, and direct radiation from the Sun is less important. The so-called hot Jupiter exoplanets, by comparison, are ten to thirty times closer to their parent stars and receive one hundred to one thousand times more radiation from their stars. Stellar X-rays, especially during flares, can cause excess heating and atmospheric changes, and over the lifetime of the planet could erode away as much as 25 percent of its mass. The star-planet interaction works both ways. For several stars with exoplanets, there is an indication that a significant increase in X-ray activity coincides with the closest approach of an exoplanet to the parent star.

The effect of X-rays from the parent star on its planets is not confined to the turbulent youth of a star. Chandra's observations of the transit of the hot Jupiter exoplanet HD 189733b around a 5-billion-year-old star similar to the Sun show that, when this exoplanet eclipsed its parent star, it produced a significantly deeper dip in soft X-rays than at optical wavelengths. This difference is attributed to a thin outer planetary atmosphere. The observations suggest that this atmosphere has a radius 1.75 times the planetary radius and is transparent at optical wavelengths but opaque to X-rays. The parent star has an anomalously high X-ray luminosity for its age, as does another planet-hosting star, CoRoT-2A.51. This observation could be evidence that the hot Jupiters in these systems in effect are keeping their parent stars young and active by transferring angular momentum from the planetary

83 | FLAME NEBULA

A composite image of a star cluster—NGC 2024—in the center of the Flame Nebula, about 1,400 light-years from Earth. X-rays from Chandra are the pink-purple dots in the center, and infrared data from Spitzer are the cloudy reds, greens, and light blues. A study of NGC 2024 and the Orion Nebula Cluster suggests that stars on their outskirts are older than stars in their central areas. NGC 2024's central stars are about two hundred thousand years old, and those on its outskirts are about 1.5 million years in age. In Orion, ages range from 1.2 million years in the middle to almost 2 million years near the edges.

84 | TARANTULA NEBULA

30 Doradus, a.k.a. the Tarantula Nebula, roughly 160,000 light-years from Earth in the Large Magellanic Cloud galaxy. In this composite, X-ray, optical, and infrared radiation detected by Chandra, Hubble, and Spitzer are blue, green, and red, respectively. One of the largest, most active star-forming regions in the Local Group of galaxies—which includes the Milky Way and Andromeda galaxies—the Tarantula is eight hundred light-years across and extremely bright. If it were the distance of the Orion Nebula (1,300 light-years), the Tarantula would span the area of sixty full Moons, and its optical light would be bright enough to cast shadows at night on Earth. Enormous stars in the Tarantula Nebula produce intense radiation and winds of searing multimillion-degree gas that carve gigantic bubbles in the surrounding cooler gas and dust. Other massive stars have raced through their evolution and exploded catastrophically as supernovas, expanding these bubbles into X-ray-brightened superbubbles. They leave behind pulsars as beacons of their former lives and expanding supernova remnants that trigger the collapse of giant clouds of dust and gas to form new generations of stars. At the center of the Tarantula Nebula lies the star cluster R136, at the intersection of three superbubbles; it is most likely the most recent star cluster to form in the Tarantula.

orbit into the stellar spin. With the dizzying pace of discovery of exoplanets, it seems certain that observations of stellar-exoplanet interactions will be one of the most exciting areas of research for Chandra in the years to come.

Pulling back for a larger view, the Chandra image of the Tarantula Nebula (figure 84), in the Large Magellanic Cloud, a nearby galaxy 160,000 light-years away, provides a tableau of the drama of star formation and evolution. The Tarantula, also known as 30 Doradus, is in one of the most active star-forming regions in our Local Group of galaxies. Massive stars are producing intense radiation and searing winds of multimillion-degree gas that carve out gigantic superbubbles in the surrounding gas. Other massive stars have raced through their evolution. They have been exploding catastrophically as supernovas, and leaving behind neutron stars and black holes. The expanding supernova remnants chemically enrich the interstellar medium of the galaxy, drive turbulence, generate cosmic rays, and trigger the collapse of giant clouds of dust and gas to form new clusters of stars. In other words, they are agents of renewal.

I witnessed an earthly example of a similar process a few years ago in a rural area of Southern California. Extremely dry Santa Ana winds had whipped the smoldering ashes of a controlled burn back to life. In a matter of hours, a full-fledged wildfire was racing through a river valley and up steep canyon slopes at speeds reaching twenty-five miles per hour. Within twenty-four hours, forty-five homes were burned to the ground and five thousand acres of chaparral, sagebrush, and avocado groves were blackened. Thankfully, no one was killed, although some residents managed to survive only by immersing themselves in a cold swimming pool.

Looking at the thick layer of ashes and the charred stumps of manzanita and live oak, it was difficult to imagine how the area would ever recover. Yet the following winter a series of light rains fell, and by springtime the hills were alive with all the colors of the rainbow in one of the most dazzling wildflower displays in my memory. The rug of gray ash had been transformed into a magic carpet of blue Canterbury bells and golden California poppies, and black limbs were sprouting green. The process of renewal had begun, and in extravagant fashion.

Renewal is a constant theme of nature, often triggered by violent events that unlock potential that has remained dormant for years—as in the wildflower seeds. Fortunately for us, many of the most dramatic examples of renewal occur not in our backyard but thousands of light-years away, where they can be safely observed with telescopes such as Chandra.

Now well into its eighteenth year of operation, Chandra has given us an unparalleled look into a realm of high-energy phenomena and fundamental physics not readily accessible with other telescopes, or with laboratories on Earth, and we have learned many cool things about the hot universe. These discoveries, while advancing our knowledge, have often revealed an underlying richness and complexity waiting to be explored. Chandra is in excellent condition, and so on the premise of "what's past is prologue," we can expect that it will continue to play a major role in this exploration.

REFERENCES

References to much of the work described in this book, and up-to-date coverage of recent developments, can be found on the Chandra X-Ray Center website, www.chandra.harvard .edu. Current and archived research papers in astrophysics are available at http://arxiv.org/ archive/astro-ph. The building of the Chandra X-Ray Observatory is discussed in detail in Wallace Tucker and Karen Tucker, *Revealing the Universe* (Cambridge, MA: Harvard University Press, 2001). A technical summary of Chandra discoveries up to 2014 is available in Harvey Tananbaum et al., "Highlights and Discoveries from the Chandra X-Ray Observatory," *Reports on Progress in Physics* 77, no. 6 (2014): 066902. Riccardo Giacconi discusses many aspects of the development of X-ray astronomy, in which he played a major role, in his autobiography, *Secrets of the Hoary Deep: A Personal History of Modern Astronomy* (Baltimore: Johns Hopkins University Press, 2008).

Introduction. Cool Stories from the Hot Universe
Einstein, Albert. 1949. *The World as I See It*. New York: Philosophical Library.
Galileo, Galilei. 1957. *The Starry Messenger* (1610). In Stillman Drake, ed. and trans., *Discoveries and Opinions of Galileo*. Garden City, NY: Doubleday.
Giacconi, Riccardo. 2008. *Secrets of the Hoary Deep: A Personal History of Modern Astronomy*. Baltimore: Johns Hopkins University Press.
Gibson, William. 2007. *Spook Country*. New York: Putnam.
Gladwell, Malcolm. 1997. "The Coolhunt." *New Yorker*, March 17: 78.
King, Stephen. 2007. "Stephen King on Who's Cool and Who's Not." *Entertainment Weekly*, www.ew.com/article/2007/11/09/stephen-king-whos-cool-and-whos-not.
Tucker, Wallace, and Karen Tucker. 2001. *Revealing the Universe: The Making of the Chandra X-Ray Observatory*. Cambridge, MA: Harvard University Press.

1. Galaxy Clusters
Abell, George. 1958. "The Distribution of Rich Clusters of Galaxies." *Astrophysical Journal Supplement* 3: 211.
Sloan Digital Sky Survey. www.sdss.org/surveys/.
Tucker, Wallace, Harvey Tananbaum, and Andrew Fabian. 2007. "Black Hole Blowback." *Scientific American* 296, no. 3 (March): 22.

2. The Evidence for Dark Matter
Bergström, Lars. 2013. "Cosmology and the Dark Matter Frontier." *Physica Scripta* 158: 14014.
Bradač, Maruša. 2008. "Revealing the Properties of Dark Matter in the Merging Cluster MACS J0025.4-1222." *Astrophysical Journal* 687: 959.
Clowe, Douglas, Anthony Gonzales, and Maxim Markevitch. 2004. "Weak-Lensing Mass Reconstruction of the Interacting Cluster 1E 0657-558: Direct Evidence for the Existence of Dark Matter." *Astrophysical Journal* 604: 596.
Conan Doyle, Arthur. 1890. *The Sign of the Four*. New York: Penguin Classics, 2001.

Fabricant, D., and P. Gorenstein. 1983. "Further Evidence for M87's Massive Dark Halo." *Astrophysical Journal* 267: 535–46.

Ford, W. Jr., V. Rubin, and M. Roberts. 1971. "A Comparison of 21-cm Radial Velocities and Optical Radial Velocities of Galaxies." *Astronomical Journal* 76: 102.

Gaskins, J. 2016. "A Review of Indirect Searches for Particle Dark Matter." *Contemporary Physics.* 2016arXiv160400014G.

Kravtsov, Andrey V., and Stefano Borgani. 2012. "Formation of Galaxy Clusters." *Annual Review of Astronomy and Astrophysics* 50: 353.

Mantz, Adam B., et al. 2014. "Cosmology and Astrophysics from Relaxed Galaxy Clusters II: Cosmological Constraints." *Monthly Notices of the Royal Astronomical Society* 440: 2077.

Milgrom, Mordecai. 1983. "A Modification of the Newtonian Dynamics as a Possible Alternative to the Hidden Mass Hypothesis." *Astrophysical Journal* 270: 365.

———. 2015. "Ultra-Diffuse Cluster Galaxies as Key to the MOND Cluster Conundrum." *Monthly Notices of the Royal Astronomical Society* 454: 3810.

Roberts, Morton. 1976. "The Rotation Curves of Galaxies." *Comments on Astrophysics* 6: 105.

Roberts, Morton, and Rots, Arnold. 1973. "Comparison of Rotation Curves of Different Galaxy Types." *Astronomy and Astrophysics* 26: 28.

Rubin, Vera. 1980. "Stars, Galaxies, Cosmos: The Past Decade, the Next Decade." *Science* 209: 63.

Rubin, Vera, and W. K. Ford Jr. 1970. "Rotation of the Andromeda Nebula from a Spectroscopic Survey of Emission Regions." *Astrophysical Journal* 159: 379.

Rubin, Vera, W. K. Ford Jr., and N. Thonnard. 1980. "Rotational Properties of 21 SC Galaxies with a Large Range of Luminosities and Radii, from NGC 4605 /R = 4kpc/ to UGC 2885 /R = 122 kpc/." *Astrophysical Journal* 238: 471.

Sanders, Jeremy, et al. 2013. "Linear Structures in the Core of the Coma Cluster of Galaxies." *Science* 341: 1365.

Spergel, David. 2015. "The Dark Side of Cosmology: Dark Matter and Dark Energy." *Science* 347: 1100.

Spergel, David, R. Flauger, and R. Raphael. 2015. "Planck Data Reconsidered." *Physical Review D* 91: 023518.

Steigman, Gary. 2007. "Primordial Nucleosynthesis in the Precision Cosmology Era." *Annual Review of Nuclear and Particle Science* 57: 463.

Strigari, L., et al. 2008. "A Common Mass Scale for Satellite Galaxies of the Milky Way." *Nature* 454: 1096.

Wolf, J., et al. 2010. "Accurate Masses for Dispersion-Supported Galaxies." *Monthly Notices of the Royal Astronomical Society* 406: 1220.

Zwicky, Fritz. 1933. "Die Rotverschiebung von extragalaktischen Nebeln." *Helvetica Physica Acta* 6: 110.

———. 1937. "On the Masses of Nebulae and Clusters of Nebulae." *Astrophysical Journal* 86: 217.

3. Cold Dark Matter

Alcock, Charles, et al. 2001. "Macho Project Limits of Black Hole Matter in the 1–30 Solar Mass Range." *Astrophysical Journal* 550: L169.

Blumenthal, George, et al. 1984. "Formation of Galaxies and Large-Scale Structure with Cold Dark Matter." *Nature* 311: 517.

Feng, J. 2010. "Dark Matter Candidates from Particle Physics and Methods of Detection." *Annual Review of Astronomy and Astrophysics* 48: 495.

Griest, K., A. Cieplak, and M. Lehner. 2013. "New Limits on Primordial Black Hole Dark Matter from an Analysis of Kepler Source Microlensing Events." *Physical Review Letters* 111: 181302.

Harvey, D., et al. 2015. "The Nongravitational Interactions of Dark Matter in Colliding Galaxy Clusters." *Science* 347: 1462.

Jee, M. James, et al. 2014. "Weighing 'El Gordo' with a Precision Scale: Hubble Space Telescope Weak-Lensing Analysis of the Merging Galaxy Cluster ACT-CL J0102-4915 at z = 0.87." *Astrophysical Journal* 785: 201.

Mantz, Adam B., et al. 2014. "Cosmology and Astrophysics from Relaxed Galaxy Clusters II: Cosmological Constraints." *Monthly Notices of the Royal Astronomical Society* 440: 2077.

Markevitch, Maxim, et al. 2004. "Direct Constraints on the Dark Matter Self-Interaction Cross Section from the Merging Galaxy Cluster 1E 0657-56." *Astrophysical Journal* 606: 819.

Morandi, A., and M. Limousin. 2012. "Triaxiality, Principal Axis Orientation and Non-Thermal Pressure in Abell 383." *Monthly Notices of the Royal Astronomical Society* 421: 3147.

Newman, A., et al. 2011. "The Dark Matter Distribution in Λ383: Evidence for a Shallow Density Cusp from Improved Lensing, Stellar Kinematic, and X-Ray Data." *Astrophysical Journal* 728: L29.

Pontzen, A., and F. Governato. 2014. "Cold Dark Matter Heats Up." *Nature* 506: 171.

Sanders, Jeremy, et al. 2013. "Linear Structures in the Core of the Coma Cluster of Galaxies." *Science* 341: 1365.

Somerville, Rachel, and Romeel Davé. 2015. "Physical Models of Galaxy Formation in a Cosmological Framework." *Annual Review of Astronomy and Astrophysics* 53: 0515.

4. Dark Matter Going Bananas

Abazajian, Kervoork, G. Fuller, and Wallace Tucker. 2001. "Direct Detection of Warm Dark Matter in the X-Ray." *Astrophysical Journal* 562: 593.

Ackermann, M., et al. 2015. "Searching for Dark Matter Annihilation from Milky Way Dwarf Spheroidal Galaxies with Six Years of Fermi Large Area Telescope Data." *Physical Review Letters* 115: 1301.

Adhikari, R., et al. 2016. "A White Paper on keV Sterile Neutrino Dark Matter." arXiv:1602.04816, High Energy Physics (hep-ph).

Anderson, Michael E., Eugene Churazov, and Joel N. Bregman. 2015. "Non-Detection of X-Ray Emission from Sterile Neutrinos in Stacked Galaxy Spectra." *Monthly Notices of the Royal Astronomical Society* 452, no. 4: 3905.

Boyarsky, Alexey, et al. 2014a. "Unidentified Line in X-Ray Spectra of the Andromeda Galaxy and Perseus Galaxy Cluster." *Physical Review Letters* 113: 1301.

———. 2014b. "Comment on 'Dark Matter Searches Going Bananas: The Contribution of Potassium (and Chlorine) to the 3.5 keV Line' by T. Jeltema and S. Profumo." arXiv:1408.4388, Cosmology and Nongalactic Astrophysics (Astro-ph.CO).

Bulbul, Esra, et al. 2014a. "Detection of an Unidentified Emission Line in the Stacked X-Ray Spectrum of Galaxy Clusters." *Astrophysical Journal* 789: 13.

———. 2014b. "Dark Matter Searches Going Bananas: The Contribution of Potassium (and Chlorine) to the 3.5 keV Line" by T. Jeltema and S. Profumo. arXiv:1409.4143, High Energy Astrophysical Phenomena (Astro-ph.HE).

———. 2014c. Quoted in "Mysterious X-Ray Signal Intrigues Astronomers." Chandra X-Ray Center, press release, June 24.

Carlson, Eric, and Stefano Profumo. 2014. "Cosmic Ray Protons in the Inner Galaxy and the Galactic Center Gamma-Ray Excess." *Physical Review D* 92: 023015.

Cline, J., and A. Frey. 2014. "Consistency of Dark Matter Interpretations of the 3.5 keV X-Ray Line." *Physical Review D* 90: 123537.

Daylan, Tansu, et al. 2014. "The Characterization of the Gamma-Ray Signal from the Central Milky Way: A Compelling Case for Annihilating Dark Matter." *Physics of the Dark Universe* 12 (June): 1. arXiv:1402.6703, High Energy Astrophysical Phenomena (Astro-ph.HE).

Iakubovskyi, Dmytro. 2015. "Observation of the New Emission Line at ~3.5 keV in X-Ray Spectra of Galaxies and Galaxy Clusters." arXiv:1510.00358, High Energy Astrophysical Phenomena (Astro-ph.HE).

Jeltema, Tesla, and Stefano Profumo. 2014a. "Dark Matter Searches Going Bananas: The Contribution of Potassium (and Chlorine) to the 3.5 keV Line." arXiv:1408.1699, High Energy Astrophysical Phenomena (Astro-ph.HE).

———. 2014b. "Reply to Two Comments on 'Dark Matter Searches Going Bananas: The Contribution of Potassium (and Chlorine) to the 3.5 keV Line.'" arXiv:1411.1759, High Energy Astrophysical Phenomena (Astro-ph.HE).

———. 2015. "Discovery of a 3.5 keV Line in the Galactic Centre and a Critical Look at the Origin of the Line across Astronomical Targets." *Monthly Notices of the Royal Astronomical Society* 450: 2143.

Livio, Mario, and Joseph Silk. 2014. "Physics: Broaden the Search for Dark Matter." *Nature* 507: 29.

Markevitch, Maxim. 2014. Quoted in "Mysterious X-Ray Signal Intrigues Astronomers." Chandra X-Ray Center press release, June 24.

Phillips, Kenneth, Barbara Sylwester, and Janusz Sylwester. 2015. "The X-Ray Line Feature at 3.5 keV in Galaxy Cluster Spectra." *Astrophysical Journal* 809: 5.

Profumo, Stefano, and Eric Carlson. 2014. "Cosmic Ray Protons in the Inner Galaxy and the Galactic Center Gamma-Ray Excess." *Physical Review D* 90: 23015.

Tamura, Takayuki, et al. 2015. "An X-Ray Spectroscopic Search for Dark Matter in the Perseus Cluster with Suzaku." *Publications of the Astronomical Society of Japan* 67: 23.

5. The Wonderful—and Fearful—Dark Side

Allen, S., et al. 2004. "Constraints on Dark Energy from Chandra Observations of the Largest Relaxed Galaxy Clusters." *Monthly Notices of the Royal Astronomical Society* 353: 457.

Anderson, Lauren, et al. 2014. "The Clustering of Galaxies in the SDSS-III Baryon Oscillation Spectroscopic Survey: Baryon Acoustic Oscillations in the Data Releases

10 and 11 Galaxy Samples." *Monthly Notices of the Royal Astronomical Society* 441, no. 1: 24–62.

Eddington, Arthur. 1978. Quoted in Horace Freeland Judson, "Annals of Science II—DNA." *New Yorker* (December 4): 132.

Einstein, Albert. 1954. *Relativity: The Special and General Theory. A Popular Exposition.* London: Methuen.

Frieman, Joshua A., Michael S. Turner, and Dragan Huterer. 2008. "Dark Energy and the Accelerating Universe." *Annual Review of Astronomy and Astrophysics* 46: 385–432.

Guth, Alan, and Paul Steinhardt. 1984. "The Inflationary Universe." *Scientific American* 250 (May): 116.

Harrison, Edward. 2000. *Cosmology: The Science of the Universe.* 2nd ed. Cambridge: Cambridge University Press.

Hubble, Edwin. 1936. *The Realm of the Nebulae.* New Haven, CT: Yale University Press.

Linde, Andrei. 1987. "Particle Physics and Inflationary Cosmology." *Physics Today* 40: 61.

Panek, Richard. 2011. *The 4% Universe: Dark Matter, Dark Energy, and the Race to Discover the Rest of Reality.* New York: Houghton Mifflin.

Perlmutter, Saul. 1999. "Measurements of Ω and Λ from 42 High-Redshift Supernovae." *Astrophysical Journal* 517: 585.

Planck Collaboration et al. 2013. "Planck 2013 Results. I. Overview of Products and Scientific Results." *Astronomy & Astrophysics* 571: 1.

Poe, Edgar Allan. 1845. "The Raven." *American Review* (February).

———. 1848. *Eureka: A Prose Poem.* New York: George P. Putnam.

Rapetti, David, et al. 2013. "A Combined Measurement of Cosmic Growth and Expansion from Clusters of Galaxies, the CMB and Galaxy Clustering." *Monthly Notices of the Royal Astronomical Society* 432: 973.

Reiss, Adam, et al. 1998. "Observational Evidence from Supernovae for an Accelerating Universe and a Cosmological Constant." *Astronomical Journal* 116: 1009.

Sagan, Carl. 1980. *Cosmos.* New York: Random House.

Schmidt, Brian, et al. 1998. "The High-Z Supernova Search: Measuring Cosmic Deceleration and Global Curvature of the Universe Using Type IA Supernovae." *Astrophysical Journal* 507: 46–63.

Spergel, David. 2015. "The Dark Side of Cosmology: Dark Matter and Dark Energy." *Science* 347: 1100.

Springel, Volker, Carlos Frenk, and Simon White. 2006. "The Large-Scale Structure of the Universe." *Nature* 440: 1137.

Turner, Michael S., and Dragan Huterer. 2007. "Cosmic Acceleration, Dark Energy, and Fundamental Physics." *Journal of the Physics Society of Japan* 76, no. 11: 111015.

Turner, Michael S., and M. White. 1997. "CDM Models with a Smooth Component." *Physics Review D* 56: 4439.

Vikhlinin, Alexey. 2008. Quoted in NASA press release, "Dark Energy Found Stifling Growth in Universe." December 16. http://chandra.harvard.edu/press/08_releases/press_121608.html.

———. 2009. "Chandra Cluster Cosmology Project III: Cosmological Parameter Constraints." *Astrophysical Journal* 692: 1060.

6. What Is Dark Energy?

Blandford, Roger. 2015. "A Century of General Relativity: Astrophysics and Cosmology." *Science* 347: 1103.

Caldwell, Robert L., and Marc Kamionkowski. 2009. "The Physics of Cosmic Acceleration." *Annual Review of Nuclear and Particle Science* 59: 397–429.

Frieman, Joshua A., Michael S. Turner, and Dragan Huterer. 2008. "Dark Energy and the Accelerating Universe." *Annual Review of Astronomy and Astrophysics* 46: 385–432.

Linde, Andrei. 2015. "A Brief History of the Multiverse." arXiv:1512.01203, High-Energy Physics—Theory (hep-th).

Rapetti, David, et al. 2010. "The Observed Growth of Massive Galaxy Clusters—III. Testing General Relativity on Cosmological Scales." *Monthly Notices of the Royal Astronomical Society* 406, no. 3: 1796–1804.

Rossi, Bruno. 1969. Personal communication.

Schmidt, Fabian, Alexey Vikhlinin, and Wayne Hu. 2009. "Cluster Constraints on f(R) Gravity." *Physical Review D* 80: 3505.

Spergel, David. 2015. "The Dark Side of Cosmology: Dark Matter and Dark Energy." *Science* 347: 1100.

7. The Cosmic Web

Bond, J. Richard, Lev Kofman, and Dmitri Pogosyan. 1996. "How Filaments of Galaxies Are Woven into the Cosmic Web." *Nature* 380: 603–06.

Cautun, Marius, et al. 2014. "Evolution of the Cosmic Web." *Monthly Notices of the Royal Astronomical Society* 441, no. 4: 2923–73.

Fang, Taotao, et al. 2010. "Confirmation of X-Ray Absorption by Warm-Hot Intergalactic Medium in the Sculptor Wall." *Astrophysical Journal* 714, no. 2: 1715–24.

Ma, C., H. Ebeling, and E. Barrett. 2009. "An X-Ray/Optical Study of the Complex Dynamics of the Core of the Massive Intermediate-Redshift Cluster MACSJ0717.5+3745." *Astrophysical Journal Letters* 693: L56.

Nicastro, F., et al. 2013. "Chandra View of the Warm-Hot Intergalactic Medium Toward 1ES 1553+113: Absorption-Line Detections and Identifications. I." *Astrophysical Journal* 769: 90.

Primack, J. 2015. "Cosmological Structure Formation." arXiv:1505.02821, Astrophysics of Galaxies (Astro-ph.GA).

Vogelsberger, Mark, et al. 2014 "Introducing the Illustris Project: Simulating the Co-Evolution of Dark and Visible Matter in the Universe." *Monthly Notices of the Royal Astronomical Society* 444: 1518.

8. Taking Pleasure in the Dimness of Stars

Abbott, B. P., et al. (LIGO Scientific Collaboration, Virgo Collaboration). 2016. "Observation of Gravitational Waves from a Binary Black Hole Merger." *Physical Review Letters* 116: 061102.

Chandrasekhar, Subramanyan. 1983. *The Mathematical Theory of Black Holes*. New York: Oxford University Press.

Coleridge, Samuel Taylor. 1798. "The Nightingale." In *The Complete Poems*, ed. William Keach. New York: Penguin Classics, 1997.

Giacconi, Riccardo. 2008. *Secrets of the Hoary Deep: A Personal History of Modern Astronomy.* Baltimore: Johns Hopkins University Press.

Hillebrandt, W., H. Janka, and E. Mueller. 2006. "How to Blow up a Star." *Scientific American* 295: 42.

Jonker, Peter, et al. 2007. "The Cold Neutron Star in the Soft X-Ray Transient 1H 1905+000." *Astrophysical Journal* 665: L147.

Lasota, J. 2008. "ADAFs, Accretion Discs and Outbursts in Compact Binaries." *New Astronomy Review* 51, no. 752.

Laycock, Silas, et al. 2015. "Revisiting the Dynamical Case for a Massive Black Hole in IC10 X-1." *Monthly Notices of the Royal Astronomical Society: Letters* 452: L31.

Liu, J., et al. 2008. "Precise Measurement of the Spin Parameter of the Stellar-Mass Black Hole M33 X-7." *Astrophysical Journal Letters* 679: L37.

McClintock, J., et al. 2007. "Chandra and Constellation-X Home in on the Event Horizon, Black Hole Spin and the Kerr Metric." *Chandra Newsletter* 14: 1.

Narayan, R., M. Garcia, and J. McClintock. 1997. "Advection-Dominated Accretion and Black Hole Event Horizons." *Astrophysical Journal Letters* 478: L97.

Narayan, R., and J. McClintock. 2014. "Observational Evidence for Black Holes." In *General Relativity and Gravitation: A Centennial Perspective.* Edited by A. Ashtekar, B. Berger, J. Isenberg, and M. A. H. MacCallum. Cambridge: Cambridge University Press.

Orosz, Jerome, et al. 2007. "A 15.65-Solar-Mass Black Hole in an Eclipsing Binary in the Nearby Spiral Galaxy M33." *Nature* 449: 8720.

Tucker, Wallace, and Riccardo Giacconi. 1985. *The X-Ray Universe.* Cambridge, MA: Harvard University Press.

Woosley, S., and A. Heger. 2015. "The Deaths of Very Massive Stars." In *Very Massive Stars in the Local Universe,* ch. 7, 199. Edited by Joris S. Vink. Astrophysics and Space Science Library 412. New York: Springer.

9. Cygnus X-1, Microquasars, and the Galactic Jet Set

Blandford, R., and R. Znajek. 1977. "Electromagnetic Extraction of Energy from Kerr Black Holes." *Monthly Notices of the Royal Astronomical Society* 179: 433.

Corbel, S., et al. 2002. "Large-Scale, Decelerating, Relativistic X-Ray Jets from the Microquasar XTE J1550-564." *Science* 298: 196.

Done, C. M. Gierliński, and A. Kubota. 2007. "Modelling the Behaviour of Accretion Flows in X-Ray Binaries: Everything You Always Wanted to Know about Accretion but Were Afraid to Ask." *Astronomy & Astrophysics Reviews* 15: 1.

Fender, R., and E. Gallo. 2014. "An Overview of Jets and Outflows in Stellar Mass Black Holes." *Space Science Reviews* 183: 323.

Giacconi, Riccardo. 2008. *Secrets of the Hoary Deep: A Personal History of Modern Astronomy.* Baltimore: Johns Hopkins University Press.

Gou, L., et al. 2011. "The Extreme Spin of the Black Hole in Cygnus X-1." *Astrophysical Journal* 742: 85.

Marshall, H., et al. 2013. "Multiwavelength Observations of the SS 433 Jets." *Astrophysical Journal* 775: 75.

McClintock, J., R. Narayan, and J. Steiner. 2014. "Black Hole Spin via Continuum Fitting and the Role of Spin in Powering Transient Jets." *Space Science Reviews* 183: 295.

Narayan, R., and J. McClintock. 2012. "Observational Evidence for a Correlation between Jet Power and Black Hole Spin." *Monthly Notices of the Royal Astronomical Society* 419: L69.

Narayan, R., J. McClintock, and A. Tchekhovskoy. 2014. "Energy Extraction from Spinning Black Holes via Relativistic Jets." *General Relativity, Cosmology and Astrophysics, Fundamental Theories of Physics* 177: 523.

Orosz, Jerome, et al. 2011. "The Mass of the Black Hole in Cygnus X-1." *Astrophysical Journal* 742: 84.

Reid, M., et al. 2011. "The Trigonometric Parallax of Cygnus X-1." *Astrophysical Journal* 742: 83.

Reynolds, C. 2014. "Measuring Black Hole Spin Using X-Ray Reflection Spectroscopy." *Space Science Reviews* 183: 277.

Soria, R., et al. 2010. "Radio Lobes and X-Ray Hot Spots in the Microquasar S26." *Monthly Notices of the Royal Astronomical Society* 409: 541.

Thorne, Kip S. 1994. *Black Holes and Time Warps: Einstein's Outrageous Legacy*. New York: Norton.

Tucker, Wallace, and Riccardo Giacconi. 1985. *The X-Ray Universe*. Cambridge, MA: Harvard University Press.

Woosley, S., and A. Heger. 2015. "The Deaths of Very Massive Stars." In *Very Massive Stars in the Local Universe*, ch. 7, 199. Edited by Joris S. Vink. Astrophysics and Space Science Library 412. New York: Springer.

10. Downtown Milky Way

Balick, B., and Brown, R. 1974. "Intense Sub-Arcsecond Structure in the Galactic Center." *Astrophysical Journal* 194: 265.

Broderick, A., A. Loeb, and R. Narayan. 2009. "The Event Horizon of Sagittarius A*." *Astrophysical Journal* 701: 1357.

Broderick, A., et al. 2015. "The Event Horizon of M87." *Astrophysical Journal* 805: 179.

Clavel, M. 2013. "Evidence of Multiple Outbursts of Sagittarius A* Revealed by Chandra." *Astronomy & Astrophysics* 558: 32.

Genzel, R., and V. Karas. 2007. "The Galactic Center." In *Black Holes from Stars to Galaxies—Across the Range of Masses*, 173. Edited by V. Karas and G. Matt. Proceedings of IAU Symposium no. 238, August 21–25, 2006, Prague, Czech Republic. Cambridge: Cambridge University Press, 2007.

Ghez, A., et al. 2008. "Measuring Distance and Properties of the Milky Way's Central Supermassive Black Hole with Stellar Orbits." *Astrophysical Journal* 689: 1044.

Inui, T., et al. 2009. "Time Variability of the Neutral Iron Lines from the Sagittarius B2 Region and Its Implication of a Past Outburst of Sagittarius A*." *Publications of the Astronomical Society of Japan* 61: 214.

Muno, M., et al. 2007. "Echoes of Multiple Outbursts of Sagittarius A* Revealed by Chandra." *Astrophysical Journal Letters* 656: L69.

Narayan, R., and J. McClintock. 2014. "Observational Evidence for Black Holes." In *General Relativity and Gravitation: A Centennial Perspective*. Edited by A. Ashtekar, B. Berger, J. Isenberg, and M. A. H. MacCallum. Cambridge: Cambridge University Press.

Nielsen, J., et al. 2013. "A Chandra/HETGS Census of X-Ray Variability from Sgr A* during 2012." *Astrophysics Journal* 774: 42.

Ovid. 8 CE. *Metamorphoses*, bk. I, 151. Translated by Frank Justus Miller. Vol. 1 (1956); vol. 2 (1958). First printing 1916, LCL. London: Heinemann.

Ponti, G., et al. 2015. "Fifteen Years of XMM-Newton and Chandra Monitoring of Sgr A: Evidence for a Recent Increase in the Bright Flaring Rate." *Monthly Notices of the Royal Astronomical Society* 454: 1525.

Schödel, R. 2016. "The Milky Way's Nuclear Star Cluster and Massive Black Hole." *Star Clusters and Black Holes in Galaxies across Cosmic Time, Proceedings of the International Astronomical Union, IAU Symposium* 312, 274.

Wang, Q. D., et al. 2013. "Dissecting X-Ray Emitting Gas around the Center of Our Galaxy." *Science* 341: 981.

11. The Secret in the Middle

Baade, Walter. 1956. Quoted in John Pfeiffer, *The Changing Universe*. London: Victor Gollancz Ltd.

———. 1965. Quoted in I. Robinson et al. *Quasi-Stellar Sources and Gravitational Collapse*. Chicago: University of Chicago Press.

Blandford, Roger, and Martin Rees. 1974. "A 'Twin-Exhaust' Model for Double Radio Sources." *Monthly Notices of the Royal Astronomical Society* 169: 395.

Fabian, Andrew. 2001. "The Energy Output of the Universe." *X-Ray Astronomy: Stellar Endpoints, AGN, and the Diffuse X-Ray Background, American Institute of Physics Conference Proceedings* 599: 93. Edited by N. White et al. Melville, NY: American Institute of Physics.

Frost, Robert. 1942. "The Secret Sits." In *The Witness Tree*. New York: Holt.

Hey, Stanley, S. J. Parson, and J. W. Phillips. 1946. "Cosmic Radiations at 5 Metres Wave-Length." *Nature* 157: 296.

Lang, Kenneth R., and Owen Gingerich. 1979. *A Source Book in Astronomy and Astrophysics, 1900–1975*. Cambridge, MA: Harvard University Press.

Lynden-Bell, Donald, and Martin Rees. 1971. "On Quasars, Dust and the Galactic Centre." *Monthly Notices of the Royal Astronomical Society* 152: 461.

Pfeiffer, John. 1956. *The Changing Universe*. London: Victor Gollancz Ltd.

Robinson, I., et al. 1965. *Quasi-Stellar Sources and Gravitational Collapse*. Chicago: University of Chicago Press.

Salpeter, Edwin. 1964. "Accretion of Interstellar Matter by Massive Objects." *Astrophysical Journal* 140: 796.

Schmidt, Maarten. 1963. "3C273: A Star-Like Object with Large Red-Shift." *Nature* 197: 1040.

Spitzer, Lyman Jr., and Walter Baade. 1951. "Stellar Populations and Collisions of Galaxies." *Astrophysical Journal* 113: 413.

Thorne, Kip S. 1994. *Black Holes and Time Warps: Einstein's Outrageous Legacy*. New York: Norton.

Zeldovich, Yakov. 1964. "The Fate of a Star and the Evolution of Gravitational Energy upon Accretion." *Soviet Physics Doklady* 9: 195.

12. Ducks Unlimited

Adams, Douglas. *Dirk Gently's Holistic Detective Agency*. New York: Simon & Schuster, 1987.

Burbidge, E. Margaret, Geoffrey Burbidge, and Kevin Prendergast. 1962. "The Rotation and Velocity Field of NGC 253." *Astrophysical Journal* 136: 339.

———. 1963. "The Rotation and Physical Conditions in the Seyfert Galaxy NGC 7469." *Astrophysical Journal* 137: 1022.

Burbidge, Geoffrey, E. Margaret Burbidge, and Allan Sandage. 1963. "Evidence for the Occurrence of Violent Events in the Nuclei of Galaxies." *Reviews of Modern Physics* 35: 94.

Chon, G., et al. 2012. "Discovery of an X-Ray Cavity Near the Radio Lobes of Cygnus A Indicating Previous AGN Activity." *Astronomy & Astrophysics* 545: L3.

Collin, Suzy. 2008. "Quasars and Galactic Nuclei, a Half-Century Agitated Story." *Albert Einstein Century International Conference. AIP Conference Proceedings* 861: 587.

Comastri, A., et al. 2015. "Mass without Radiation: Heavily Obscured AGNs, the X-Ray Background, and the Black Hole Mass Density." *Astronomy & Astrophysics* 574: L10.

Fabian, Andrew. 2001. "The Energy Output of the Universe." *X-Ray Astronomy: Stellar Endpoints, AGN, and the Diffuse X-Ray Background, American Institute of Physics Conference Proceedings* 599, 93. Edited by N. White et al. Melville, NY: American Institute of Physics.

Giommi, P., and P. Padovani. 2015. "A Simplified View of Blazars: Contribution to the X-Ray and γ-Ray Extragalactic Backgrounds." *Monthly Notices of the Royal Astronomical Society* 450: 240.

Schmidt, Maarten. 1963. "3C273: A Star-Like Object with Large Red-Shift." *Nature* 197: 1040.

Seyfert, Carl. 1943. "Nuclear Emission in Spiral Nebulas." *Astrophysical Journal* 97: 28.

Ueda, Yoshihiro. 2015. "Review: Cosmological Evolution of Supermassive Black Holes in Galactic Centers Unveiled by Hard X-Ray Observations." *Proceedings of the Japan Academy, Ser. B: Physical and Biological Sciences* 91, no. 5: 175–92.

Xue, Y. Q., et al. 2012. "The Chandra Deep Field-South Survey: 4 Ms Source Catalogs." *Astrophysical Journal* 758: 129.

13. The Origin and Growth of Supermassive Black Holes

Armus, L., et al. 2009. "GOALS: The Great Observatories All-Sky LIRG Survey." *Publications of the Astronomical Society of the Pacific* 121: 559.

Baade, Walter, and Edwin Hubble. 1939. "The New Stellar Systems in Sculptor and Fornax." *Publications of the Astronomical Society of the Pacific* 51: 40.

Baldassare, Vivienne. 2015. "Oxymoronic Black Hole Provides Clues to Growth." NASA press release, August 11. http://chandra.harvard.edu/press/15_releases/press_081115 .html.

Baldassare, Vivienne, et al. 2015. "A ~50,000 Solar Mass Black Hole in the Nucleus of RGG 118." *Astrophysical Journal* 809: L14.

Fabbiano, Giuseppina. "NASA's Chandra Finds Nearest Pair of Supermassive Black Holes." NASA press release, August 31, 2011. http://chandra.harvard.edu/press/11 _releases/press_083111.html.

Fabbiano, Giuseppina, et al. 2011. "A Close Nuclear Black-Hole Pair in the Spiral Galaxy NGC3393." *Nature* 477: 431.

Green, Paul, et al. 2013. "SDSS J1254+0846: A Binary Quasar Caught in the Act of Merging." *Astrophysical Journal* 710: 1578.

Pacucci, Fabio, et al. 2015. "Shining in the Dark: The Spectral Evolution of the First Black Hole." *Monthly Notices of the Royal Astronomical Society* 454: 377.

Page, M., et al. 2014. "X-Rays from the Redshift 7.1 Quasar ULAS J1120+0641." *Monthly Notices of the Royal Astronomical Society* 440: L91.

Volonteri, M., and J. Bellovary. 2012. "Black Holes in the Early Universe." *Reports on Progress in Physics* 75: 12490.

Wu, X., et al. 2015. "An Ultraluminous Quasar with a Twelve-Billion-Solar-Mass Black Hole at Redshift 6.30." *Nature* 518: 512.

Xue, Y. Q., et al. 2012. "The Chandra Deep Field-South Survey: 4 Ms Source Catalogs." *Astrophysical Journal* 758: 129.

Zwicky, Fritz, and Magrit Zwicky. 1971. *Catalogue of Selected Compact Galaxies and of Post-Eruptive Galaxies.* Zurich: L-Spreich.

14. Green Black Holes

Allen, Steve, et al. 2006. "The Relation between Accretion Rate and Jet Power in X-Ray Luminous Elliptical Galaxies." *Monthly Notices of the Royal Astronomical Society* 372: 21.

Edmonds, Peter. 2005. *Chandra Chronicles.* January 5.

MacLeod, C., et al. 2015. "A Consistent Picture Emerges: A Compact X-Ray Continuum Emission Region in the Gravitationally Lensed Quasar SDSS J0924+0219." *Astrophysical Journal* 806: 258.

Marrone, D., et al. 2007. "An Unambiguous Detection of Faraday Rotation in Sagittarius A*." *Astrophysical Journal* 654: L57.

Miller, J., et al. 2015. "Flows of X-Ray Gas Reveal the Disruption of a Star by a Massive Black Hole." *Nature* 526: 542.

NASA. 2004. "Giant Black Hole Rips Apart Star." Press release. February 18.

———. 2006. "NASA's Chandra Finds Black Holes Are 'Green.'" Press release. April 24.

Reynolds, M. 2014. "A Rapidly Spinning Black Hole Powers the Einstein Cross." *Astrophysical Journal* 792: L19.

Russell, H., et al. 2015. "Inside the Bondi Radius of M87." *Monthly Notices of the Royal Astronomical Society* 451: 588.

Wang, H., and D. Merritt. 2004. "Revised Rates of Stellar Disruption in Galactic Nuclei." *Astrophysical Journal* 600: 149.

Wang, Q. E., et al. 2013. "Dissecting X-Ray-Emitting Gas around the Center of Our Galaxy." *Science* 341: 98.

Wong, K., et al. 2011. "Resolving the Bondi Accretion Flow toward the Supermassive Black Hole of NGC 3115 with Chandra." *Astrophysical Journal* 736: L23.

———. 2014. "The Megasecond Chandra X-Ray Visionary Project Observation of NGC 3115: Witnessing the Flow of Hot Gas within the Bondi Radius." *Astrophysical Journal* 780: 9.

15. Black Hole Feedback

Mahdavi, A., N. Trentham, and R. Tully. 2005. "The NGC 5846 Group: Dynamics and the Luminosity Function to $M_R = -12$." *Astronomical Journal* 130: 1502.

Main, R., et al. 2015. "AGN Feedback, Host Halo Mass and Central Cooling Time: Implications for Galaxy Formation Efficiency and $M_{BH}-\sigma$." arXiv:1510.07046, Astrophysics of Galaxies (Astro-ph.GA).

McNamara, Brian, and P. Nulsen. 2007. "Heating Hot Atmospheres with Active Galactic Nuclei." *Annual Review of Astronomy and Astrophysics* 45: 117.

———. 2012. "Mechanical Feedback from Active Galactic Nuclei in Galaxies, Groups and Clusters." *New Journal of Physics* 14: 055023.

Nulsen, P., and Brian McNamara. 2013. "AGN Feedback in Clusters: Shock and Sound Heating." *Astronomische Nachrichten* 334, nos. 4–5: 386.

Randall, S., et al. 2015. "A Very Deep Chandra Observation of the Galaxy Group NGC 5813 AGN Shocks, Feedback, and Outburst History." *Astrophysical Journal* 805: 112.

Tananbaum, Harvey, et al. 2014. "Highlights and Discoveries from the Chandra X-Ray Observatory." *Reports on Progress in Physics* 77, no. 6: 066902.

Tucker, Wallace, Harvey Tananbaum, and Andrew Fabian. 2007. "Black Hole Blowback." *Scientific American* 296 (March): 22.

16. Going Not Gentle into That Good Night

Baade, Walter. 1945. "B Cassiopeia as a Supernova of Type I." *Astrophysical Journal* 102: 309.

Baade, Walter, and Fritz Zwicky. 1934. "On Super-Novae." *Proceedings of the National Academy of Sciences* 20: 254.

Burkey, M., et al. 2013. "X-Ray Emission from Strongly Asymmetric Circumstellar Material in the Remnant of Kepler's Supernova." *Astrophysical Journal* 764: 63.

Cappellaro, E., et al. 2015. "Supernova Rates from the SUDARE VST-OmegaCAM Search. I. Rates per Unit Volume." *Astronomy & Astrophysics* 584: 62.

Chiotellis, A, K. Schure, and J. Vink. 2012. "The Imprint of a Symbiotic Binary Progenitor on the Properties of Kepler's Supernova Remnant." *Astronomy & Astrophysics* 537: 139.

Duyvendak, J. J. L. 1942. "Further Data Bearing on the Identification of the Crab Nebula with the Supernova of 1054 A.D. Part I. The Ancient Chronicles." *Proceedings of the Astronomical Society of the Pacific* 54: 91.

Eddington, Arthur. 1935. "Relativistic Degeneracy." *Observatory* 58: 37.

Ferris, Timothy. 1988. *Coming of Age in the Milky Way*. New York: William Morrow.

Gorenstein, Paul, E. Kellogg, and H. Gursky. 1970. "X-Ray Characteristics of Three Supernova Remnants." *Astrophysical Journal* 160: 199.

Hanbury-Brown, R., and C. Hazard. 1952. "Radio-Frequency Radiation from Tycho Brahe's Supernova (A.D. 1572)." *Nature* 170: 364.

Hillebrandt, W., H. Janka, and E. Muller. 2006. "How to Blow Up a Star." *Scientific American* (October): 42–49.

Hillebrandt, W., and F. Röpke. 2010. "Modeling Type Ia Supernovae." *New Astronomy Reviews* 54: 201.

Katsuda, S., et al. 2015. "Kepler's Supernova: An Overluminous Type Ia Event Interacting with a Massive Circumstellar Medium at a Very Late Phase." *Astrophysical Journal* 808: 49.

Kerzendorf, W., et al. 2013. "A High-Resolution Spectroscopic Search for the Remaining Donor for Tycho's Supernova." *Astrophysical Journal* 774: 99.

———. 2014. "A Reconnaissance of the Possible Donor Stars to the Kepler Supernova." *Astrophysical Journal* 782: 27.

Krause, Oliver, et al. 2008. "Tycho Brahe's 1572 Supernova as a Standard Type Ia as Revealed by Its Light-Echo Spectrum." *Nature* 456: 617.

Li, W., et al. 2011. "Nearby Supernova Rates from the Lick Observatory Supernova Search— III. The Rate-Size Relation, and the Rates as a Function of Galaxy Hubble Type and Colour." *Monthly Notices of the Royal Astronomical Society* 412: 1473L.

Miller, A. 2005 *Empire of the Stars: Obsession, Friendship and Betrayal in the Quest for Black Holes.* New York: Houghton Mifflin.

Minkowski, Rudolph. 1941. "Spectra of Supernovae." *Publications of the Astronomical Society of the Pacific* 53: 224.

Thomas, Dylan. 1952. "Do Not Go Gentle into That Good Night" (1951). In *In Country Sleep, and Other Poems.* New York: New Directions.

Wang, B., and Z. Han. 2012. "Progenitors of Type Ia Supernovae." *New Astronomy Reviews* 56: 122.

Whelan, John, and Icko Iben Jr. 1973. "Binaries and Supernovae of Type I." *Astrophysical Journal* 186: 1007.

17. Core Collapse

Delaney, T., et al. 2010. "The Three-Dimensional Structure of Cassiopeia A." *Astrophysical Journal* 725: 2038.

Delaney, T., and J. Satterfield. 2013. "The Proper Motion of the Neutron Star in Cassiopeia A." arXiv:1307.3539, High Energy Astrophysical Phenomena (Astro-ph.HE).

Helder, E., et al. 2013. "Chandra Observations of SN 1987A: The Soft X-Ray Light Curve Revisited." *Astrophysical Journal* 764: 11.

Ho, W., et al. 2015. "Tests of the Nuclear Equation of State and Superfluid and Superconducting Gaps Using the Cassiopeia A Neutron Star." *Physical Review C* 91: 5806.

Ho, W., and C. Heinke. 2009. "A Neutron Star with a Carbon Atmosphere in the Cassiopeia A Supernova Remnant." *Nature* 462: 71.

Hwang, U., and M. Laming. 2012. "A Chandra X-Ray Survey of Ejecta in the Cassiopeia A Supernova Remnant." *Astrophysical Journal* 746: 130.

Janka, H. 2012. "Explosion Mechanisms of Core-Collapse Supernovae." *Annual Review of Nuclear and Particle Science* 62: 407.

Lopez, L., et al. 2013. "The Galactic Supernova Remnant W49B Likely Originates from a Jet-Driven, Core-Collapse Explosion." *Astrophysical Journal* 764: 50L.

Park, S., et al. 2007. "A Half-Megasecond Chandra Observation of the Oxygen-Rich Supernova Remnant G292.0+1.8." *Astrophysical Journal* 670: L21.

Sukhbold, T., et al. 2016. "Core-Collapse Supernovae from 9 to 120 Solar Masses Based on Neutrino-Powered Explosions." *Astrophysical Journal* 821: 38.

Vink, Joris. 2012. "Supernova Remnants: The X-Ray Perspective." *Astronomy & Astrophysics Review* 20: 49.

Woosley, S., and A. Heger. 2015. "The Deaths of Very Massive Stars." In *Very Massive Stars in the Local Universe*, ch. 7, 199. Edited by Joris S. Vink. Astrophysics and Space Science Library 412. New York: Springer.

Woosley, S., and T. Janka. 2005. "The Physics of Core-Collapse Supernovae." *Nature Physics* 1: 147.

Yeats, William B. 1920. "The Second Coming." *Dial* (Chicago). November.

18. The Crab and Other Pulsar Wind Nebulas

Baade, Walter, and Fritz Zwicky. 1934a. "On Super-Novae." *Proceedings of the National Academy of Sciences* 20, no. 5: 254.

———. 1934b. "Cosmic Rays from Super-Novae." *Proceedings of the National Academy of Sciences* 20: 259.

Bowyer, S., et al. 1964. "Lunar Occultation of X-Ray Emission from the Crab Nebula." *Science* 146: 912.

Bühler, R., and Roger Blandford. 2014. "The Surprising Crab Pulsar and Its Nebula: A Review." *Reports on Progress in Physics* 77: 6901.

Duyvendak, J. 1942. "Further Data Bearing on the Identification of the Crab Nebula with the Supernova of 1054 A.D. Part I. The Ancient Chronicles." *Proceedings of the Astronomical Society of the Pacific* 54: 91.

Gaensler, B., and P. Slane. 2006. "The Evolution and Structure of Pulsar Wind Nebulae." *Annual Review of Astronomy and Astrophysics* 44: 17.

Gaensler, B., et al. 2004. "The Mouse That Soared: High-Resolution X-Ray Imaging of the Pulsar-Powered Bow Shock G359.23-0.822004." *Astrophysical Journal* 616: 383.

Hester, J. 2008. "The Crab Nebula: An Astrophysical Chimera." *Annual Review of Astronomy and Astrophysics* 46: 127.

Kargaltsev, O., et al. 2015. "Pulsar-Wind Nebulae: Recent Progress in Observations and Theory." *Space Science Reviews* 191: 391.

Kargaltsev, O., and G. Pavlov. 2008. "Pulsar Wind Nebulae in the Chandra Era." *40 Years of Pulsars: Millisecond Pulsars, Magnetars and More. AIP Conference Proceedings* 983: 171.

Lyutikov, M., and A. Lazarian. 2013. "Topics in Microphysics of Relativistic Plasmas." *Space Science Reviews* 178: 459.

Mayall, Nicholas U. 1962. "The Story of the Crab Nebula." *Science* 137: 91.

Mayall, Nicholas U., and Jan H. Oort. 1942. "Further Data Bearing on the Identification of the Crab Nebula with the Supernova of 1054 A.D. Part II. The Astronomical Aspects." *Publications of the Astronomical Society of the Pacific* 54, no. 318: 95.

Minkowski, Rudolph. 1942. "The Crab Nebula." *Astrophysical Journal* 96: 199.

Pacini, Franco. 1967. "Energy Emission from a Neutron Star." *Nature* 216: 567.

Staeilin, D., and E. Reifenstein. 1968. "Pulsating Radio Sources near the Crab Nebula." *Science* 162: 1481.

Stappers, B., et al. 2003. "An X-Ray Nebula Associated with the Millisecond Pulsar B1857+20." *Science* 299: 1372.

Yang, H., and R. Chevalier. 2015. "Evolution of the Crab Nebula in a Low Energy Supernova." *Astrophysical Journal* 806: 153.

Yuan, Y., and Roger Blandford. 2015. "On the Implications of Recent Observations of the Inner Knot in the Crab Nebula." *Monthly Notices of the Royal Astronomical Society* 454: 275.

19. A Thin Cosmic Rain: Particles from Outer Space

Blandford, Roger, P. Simeon, and Y. Yuan. 2014. "Cosmic Ray Origins: An Introduction." *Nuclear Physics B—Proceedings Supplements* 256: 9

Blasi, P. 2013. "The Origin of Galactic Cosmic Rays." *Astronomy and Astrophysics Review* 21: 70.

Eriksen, K., et al. 2011. "Evidence for Particle Acceleration to the Knee of the Cosmic Ray Spectrum in Tycho's Supernova Remnant." *Astrophysical Journal* 728: L28.

Ferrier, K., J. Kirchner, and R. Finkel. 2008. "Cosmogenic-Based Physical and Chemical Denudation Rates in the Idaho Batholith." *Geochimica et Cosmochimica Acta* 72: 266.

Friedlander, M. 1989. *Cosmic Rays.* Cambridge, MA: Harvard University Press.

Ginzburg, Vitaly, and Sergei Syrovatskii. 1964. *The Origin of Cosmic Rays.* New York: Macmillan.

Granger, Darryl, N. Lifton, and J. Willenbring. 2013. "A Cosmic Trip: 25 Years of Cosmogenic Nuclides in Geology." *Geological Society of America Bulletin* 125: 1379.

Greensfelder, L. 2002. "Subtleties of Sand Reveal How Mountains Crumble." *Science* 295: 256.

Hess, Victor. 1936. "Nobel Lecture: Unsolved Problems in Physics: Tasks for the Immediate Future in Cosmic Ray Studies." Nobel Media AB 2014. www.nobelprize.org/nobel_prizes/physics/laureates/1936/hess-lecture.html. April 8, 2016.

Wang, F., et al. 2015. "Beryllium-10 Concentrations in the Hyper-Arid Soils in the Atacama Desert, Chile: Implications for Arid Soil Formation Rates and El Niño Driven Changes in Pliocene Precipitation." *Geochimica et Cosmochimica Acta* 160: 227.

Warren, J., et al. 2005. "Cosmic-Ray Acceleration at the Forward Shock in Tycho's Supernova Remnant: Evidence from Chandra X-Ray Observations." *Astrophysical Journal* 674: 376.

20. Elemental Change

Burbidge, E. Margaret, Geoffrey R. Burbidge, William Fowler, and Fred Hoyle. 1957. "Synthesis of the Elements in Stars." *Reviews of Modern Physics* 29: 547.

Delaney, T., et al. 2010. "The Three-Dimensional Structure of Cassiopeia A." *Astrophysical Journal* 725: 2038.

Fabbiano, Giuseppina, et al. 2004. "X-Raying Chemical Evolution and Galaxy Formation in the Antennae." *Astrophysical Journal* 605: L21.

Fang, Taotao, et al. 2010. "Confirmation of X-Ray Absorption by Warm-Hot Intergalactic Medium in the Sculptor Wall." *Astrophysical Journal* 714, no. 2: 1715–24.

Hwang, U., and M. Laming. 2012. "A Chandra X-Ray Survey of Ejecta in the Cassiopeia A Supernova Remnant." *Astrophysical Journal* 746: 130.

Kirkpatrick, C., and B. McNamara. 2015. "Hot Outflows in Galaxy Clusters." *Monthly Notices of the Royal Astronomical Society* 452: 4361.

Strickland, D., and T. Heckman. 2009. "Supernova Feedback Efficiency and Mass Loading in the Starburst and Galactic Superwind Exemplar M82." *Astrophysical Journal* 697: 230.

Tananbaum, Harvey, et al. 2014. "Highlights and Discoveries from the Chandra X-Ray Observatory." *Reports on Progress in Physics* 72: 066902.

21. Cosmic Renewal

Getman, K., et al. 2014a. "Core-Halo Age Gradients and Star Formation in the Orion Nebula and NGC 2024 Young Stellar Clusters." *Astrophysical Journal* 787: 108.

———. 2014b. "Age Gradients in the Stellar Populations of Massive Star Forming Regions Based on a New Stellar Chronometer." *Astrophysical Journal* 787: 108.

Poppenhager, K., et al. 2013. "Transit Observations of the Hot Jupiter HD 189733b at X-Ray Wavelengths." *Astrophysical Journal* 773: 62.

Rivilla, V., et al. 2013. "The Role of Low-Mass Star Clusters in Massive Star Formation: The Orion Case." *Astronomy & Astrophysics* 554: 48.

Townsley, L., et al. 2006. "A Chandra ACIS Study of 30 Doradus. I. Superbubbles and Supernova Remnants." *Astronomical Journal* 131: 2140.

———. 2014. "The Massive Star-Forming Regions Omnibus X-Ray Catalog." *Astrophysical Journal Supplement* 213: 1.

Wolk, S., et al. 2011. "The Chandra Carina Complex Project View of Trumpler 16." *Astrophysical Journal Supplement* 194: 1.

ACKNOWLEDGMENTS

I would first like to acknowledge that I have only scratched the surface of what has been accomplished by the thousands of people who built Chandra, one of the most successful telescopes in the history of astronomy; by those who have operated it and managed its output so effectively; and by the even larger community of scientists who have used it so creatively. As of December 2015, more than 6,000 scientific papers using Chandra data had been published, and although Chandra's press scientist, Peter Edmonds, and public affairs specialist, Megan Watzke, do an excellent job in keeping up with the high-profile news from their telescope, many other important stories remain to be told.

Chandra exists because of the vision of Riccardo Giacconi and the leadership of Harvey Tananbaum, who by almost any measure was one of the most outstanding directors ever of a major astrophysical observatory. He became director in 1991, eight years before the telescope was renamed from the AXAF (Advanced X-Ray Astrophysics Facility) to Chandra upon its successful launch in 1999, and he remained in that position until 2014, longer than any other director of a world-class telescope. On a personal level, I thank Harvey and his successor, Belinda Wilkes, for offering me the opportunity to participate in a small and unconventional way in Chandra's journey. Harvey has been a valued friend and colleague for longer than either of us care to mention. My work for Chandra has been primarily with the director's office and with the core members of the public information team: Kathleen Lestition, Peter Edmonds, Megan Watzke, and Kimberly Kowal Arcand. It has been a pleasure to work with such a congenial, resourceful, and extraordinarily talented group. Joseph DePasquale, Kristin DiVona, April Hobart, John Little, Khajag Mgrdichian, Kayren Phillips, Aldo Solares, and Melissa Weiss also deserve recognition for their outstanding work as part of the Web, multimedia, imaging, and support team.

I thank Jeff McClintock, Jim Renshaw, Kerry Renshaw, and Stuart Tucker for carefully reading the manuscript and making numerous useful suggestions. I also thank Laura Harger and Gregory McNamee for their excellent editorial work; Carolyn Gleason, Smithsonian Books' director; Christina Wiginton, managing editor; and Matt Litts, marketing director. Finally, I thank my wife, Beverly, for her love, friendship, encouragement, and forbearance during the time it took to complete this book.

INDEX

Page numbers in italics indicate illustrations and captions.

CHANDRA'S COSMOS

160, *167*, 168, 179, *180*,
181–83, *184*, 185–86, *188*,
189–91, *192*, 193–96, *197*,
199, *216–17*, 226, 227, 236
Newton, Isaac, 15, 161; law
of gravity, 36, 46, 71, 109;
laws of motion, 15, 46
NGC 2024 star cluster,
232–33
NGC 3115 galaxy, 133
NGC 3393 galaxy, *120*, 127
NGC 5813 galaxy group, *147*,
148–49, 150
NGC 6240 galaxy, 127–28,
128–29
NGC 7793 galaxy, 87
"The Nightingale"
(Coleridge), 71
Norma galaxy cluster, 219
normal matter, 3, *20–21*, 30,
32, 53
Novikov, Igor, 109

Omega quantity, 49–50, 57
optical spectrums, 11
optical telescopes: Apache
Point Observatory, 11; and
Galactic Center, *94–95*;
hot gas undetectable
with, 12, 17, *143*; Kitt Peak
National Observatory, 125;
Magellan Space Telescope,
20, *126*; and mass of
dark star, 74; Palomar
telescope and observatory,
103–06, 123, 125, *187*;
revealing only a portion
of universe, 1; and spiral
galaxy NGC 3393, 127;
and supermassive black
hole population, 118; and
triple ring system, 181–82;
and X-radiation of dark,
massive objects, 110
Orion Nebula Cluster, *224*,
228, 230–33, 235
Ovid, 91, 92, 225

Pacini, Franco, 189, 191, 193
Palomar telescope and
observatory (California),
103–06, 123, 125, *187*
Parson, S. J., 102
Perlmutter, Saul, 50
Perseus galaxy cluster, *34*, 37,
142, 144–45
Peterson, John, 146
Phillips, J. W., 102
Phoenix (SPT-CLJ2344-4243)
galaxy cluster, 150, *154–55*
Pictor A galaxy, *72–73*
Planck telescope (ESA),
18–19, 19, 51
planetary nebula, 227
Poe, Edgar Allan, 43, 44
Pogosyan, Dmitri, 61
pregalactic clumps, 29. *See
also* cold dark matter
Prendergast, Kevin, 113
pre-supernova star, 172, *178*,
181, 183
Profumo, Stefano, 38–39
protostar stage of star
formation, 226, 227, 228
pulsars and pulsar wind
nebulas, *184*, 185,
189–201, 235; Black
Widow (B1957+20),
199–201, *200*; Crab
Nebula, *93*, 159–60, *188*,
189–201, *192*, *196*; Mouse,
195–99, *198*; Vela, *197*

quantum mechanics, 36, 56,
168; Fowler's new theory
of, 168
quasars, 64, *64*, 65, 81–89,
107–10, 114–15, 136,
209; binary, 125–27,
126; Cloverleaf, *140–41*;
microquasars, 81–89; RX
J1131-1231, 137–38; S26,
87–88; in Sculptor Wall,
66–67; seen in Hubble
images, 64, 124; Type 2, 115

quasi-stellar radio sources.
See quasars
"Quasi-Stellar Sources and
Gravitational Collapse"
symposium, 107

R136 star cluster, *234–35*
radiation absorption
"barcode," 35–36, 133
Radio Arc, 96
radio clouds, *72–73*
radio sources, 103, 105–06,
107; 3C273, 106–07; 3C48,
106; in Crab Nebula,
191–92; Cygnus A, 102,
105, 106; Sagittarius A*, 96
ram pressure, *218–19*
"The Raven" (Poe), 43
Reber, Grote, 102
red dwarf stars, 227
red giant supernova trigger
mechanism, 165
redshift, 62, 104, 106–07
Rees, Martin, 110
Reiss, Adam, 50
Retherford, Robert, 56
Reynolds, Christopher, 131,
136
Roentgen, Wilhelm, 82, 203
ROSAT X-ray satellite
(Roentgensatellite), *vi*, 58,
132, 144
Rossi X-Ray Timing Explorer
(RXTE), 85
RX J1131-1231 quasar, 137–38
Ryle, Martin, 102, 103

S26 microquasar, 87–88
Sagan, Carl, 50
Sagittarius A* (Sgr A*), 90,
93–94, 96–99, 133
Salpeter, Edwin, 109
Sandage, Allan, 106, 114
Schmidt, Brian, 50
Schmidt, Fabian, 58
Schmidt, Maarten, 106–07,
114

supernovas, 4, 12, 168, *170*, *178, 180*; of 1006 CE, 160, *206–07*; of 1054 CE (Crab Nebula), *92–93*, 159–60, *188*, 189–201, *192, 196*; of 1572 (Tycho's), *158*, 161–62, 169, *172*, 207–09, *210–11*
supershells, *227*
supersymmetry, 27
Swift satellite (NASA), 97
Syrovatskii, Sergei, 208

Tananbaum, Harvey, 239, 255
Tarantula Nebula (30 Doradus), *ii–iii, 234–35*
Taurus constellation, *188*
Teller, Edward, 109
theory of general relativity (Einstein), 15, 21, 47, 55–56, 58–59
theory of gravity (Einstein), 43, 46, *54*, 106
Thomas, Dylan, 159
tidal disruption, 132–33, *134–35*
torus: around black hole, 114, *116, 117*; around neutron star, 194
Treister, Ezekiel, 124
triple ring system, *176*, 181–82
Trumpler 16 star cluster, 229
Turner, Michael, 50
Tycho supernova remnant, *158*, 161–62, 169, *172*, 207–09; X-ray stripes in, *210–11*
Type 1 AGNs, *119*
Type 2 AGNs, *119*
Type Ia supernova, 44, 160, 162–63, *165, 166, 170*,

175, *227*; CCD cameras surveying, 50; Chandra observations of, 172; characteristics of, 169; G1.9+0.3 remnant, 175. *See also* supernovas; Tycho supernova remnant
Type Ib supernova, 162
Type Ic supernova, 162
Type II supernova, 162, *170*, 227

Uhuru X-ray satellite, 81, 84
UltraVista Survey (ESO), 12
universe: energy distribution of, *53*; rate of expansion of, 48, 175
Uranus, 15, 16

vacuum energy, 56–57
Very Large Array, *108*
Very Large Telescope (ESO), *42, 54*, 87, *118*
Vikhlinin, Alexey, 52–53, 58
Virgo galaxy cluster, 16, 145, 150
Vogelsberger, Mark, 62
Vulcan (theoretical planet), 15

W44 supernova remnant (G34.7-0.4), 222–23
W49B supernova remnant, 186, *187*
warm dark matter, 29
warm-hot intergalactic medium (WHIM), 64, 65, *66–67*
weakly interacting massive particles (WIMPs), 28, 40
Whelan, John, 163
WHIM. See warm-hot intergalactic medium

white dwarf stars, 4, 15, 106, 160, 163–68, *165, 167*, 172, 175, 179, 181, 185, 190, 193, 226, *227*; merging, as supernova trigger mechanism, *166*; Sirius B, *171, 173*
Wilkes, Belinda, 255
Wilkinson Microwave Anisotropy Probe (WMAP), 19, 51
WIMPs. *See* weakly interacting massive particles
Wolf-Rayet type star, 164

XMM-Newton observatory (ESA), 37–39, 65, *66–67*, 78, 82, 97–98, 132, 137, 144, 146
X-rays: and astronomy, 2, 223, 228, 239; from binary systems, 74–78, 82, 91, 96, 186; flares, 132, 224; fluorescence, 82, *83*, 98; photons, 2, 82, *83*; and supernovas, 76; X-ray-emitting wind, 132–33; X-ray telescopes, 2, 12, 17, 46, 51, 58, 74, 82–84, 91, 97, *118*, 182, 205, 208. *See also* Giacconi, Riccardo
XTE J1550 black hole system, 85, 87

Yeats, William Butler, 177

Zeldovich, Yakov, 109
Znajek, Roman, 81, 85
Zwicky, Fritz, 16–17, 122–23, 161–62, 189, 208

PHOTOGRAPHY AND
ILLUSTRATION CREDITS

Acronyms

AAO	Anglo-Australian Observatory
ALMA	Atacama Large Millimeter/Submillimeter Array
ASTRON	Netherlands Foundation Research in Astronomy
ASU	Arizona State University
ATCA	Australia Telescope Compact Array
ATNF	Australia Telescope National Facility
AUI	Associated Universities, Inc.
AURA	Associated Universities for Research in Astronomy
CfA	Harvard-Smithsonian Center for Astrophysics
CSIRO	Commonwealth Scientific and Industrial Research Organisation
CXC	Chandra X-Ray Center
DRAO	Dominion Radio Astrophysical Observatory
ESA	European Space Agency
ESO WFI	European Space Observatory Wide Field Imager
IfA	Institute for Astronomy, University of Hawaii
IoA	Institute of Astronomy, Cambridge University
JHU	Johns Hopkins University
JPL	Jet Propulsion Laboratory
MIT	Massachusetts Institute of Technology
MPE	Max Planck Institute for Extraterrestrial Physics
MYStIX	Massive Young Star-Forming Complex Study in Infrared and X-Ray
NAOJ	National Astronomical Observatory of Japan
NASA	National Aeronautics and Space Administration
NCSU	North Carolina State University
NRAO	National Radio Astronomy Observatory
NRC	National Research Council
NSF	National Science Foundation
PSU	Penn State University
ROSAT	Roentgensatellite
SAO	Smithsonian Astrophysical Observatory

SDSS (I, II, III)	Sloan Digital Sky Surveys
SDSU	San Diego State University
STScI	Space Telescope Science Institute
UAH	University of Alabama, Huntsville
UltraVISTA	Ultra-Deep Survey, Vista Telescope
VLA	Karl G. Jansky Very Large Array
VLT	Very Large Telescope, European Southern Observatory

Credits

Frontispiece, figure 84: X-ray: NASA/CXC/PSU/Leisa Townsley et al.; optical: NASA/STScI; infrared: NASA/JPL/PSU/Leisa Townsley et al. Figure 1: X-ray: NASA/CXC/Bryan Gaensler et al.; X-ray: NASA/ROSAT/I. Asaoka and B. Aschenbach; radio wide: NRC/DRAO/D. Leahy; radio detail: NRAO/VLA; optical: Digitized Sky Survey (DSS). Figure 2: NASA/CXC/SAO/Rutgers University/John Hughes. Figure 3: Illustris Collaboration. Figure 4: X-ray: NASA/CXC/MIT/En-Hsin Peng et al.; optical: NASA/STScI. Figure 5: Jeremy Tinker/SDSS-III collaboration. Figure 6: X-ray: NASA/CXC/Université de Paris/T. Wang et al.; infrared: ESO WFI/UltraVISTA; radio: ESO WFI/NAOJ/NRAO/ALMA. Figure 7: X-ray: NASA/CXC/MPE/Jeremy Sanders et al.; optical: SDSS. Figure 8: ESA and Planck Collaboration. Figure 9: X-ray: NASA/CXC/CfA/Maxim Markevitch et al.; optical: NASA/STScI/Magellan Telescope/University of Arizona/Douglas Clowe et al.; lensing map: NASA/STScI/ESO WFI/Magellan Telescope/University of Arizona/Douglas Clowe et al. Figure 10: X-ray: NASA/CXC/Stanford University/Steve Allen; optical/lensing: NASA/STScI/University of California, Santa Barbara/Maruša Bradač. Figure 11: NASA/ESA/Myungkook James Jee et al. Figure 12: NASA/MIT/John Arabadjis et al. Figure 13: NASA/CXC/SAO/Esra Bulbul et al. Figure 14: X-ray: NASA/CXC/Boston University/Elizabeth Blanton; optical: ESO WFI/VLT. Figure 15: NASA/CXC/Kristin Divona. Figure 16: X-ray: NASA/CXC/California Institute of Technology/A. Newman et al./Tel Aviv/Andrea Morandi and Marceau Limousin; optical: NASA/STScI, ESO WFI/VLT, SDSS. Figure 17: X-ray: NASA/CXC/IfA/Chen-Juin Ma et al.; optical: NASA/STScI/IfA/Chen-Juin Ma et al. Figures 18, 21: NASA/CXC/April Hobart. Figure 19: illustration: NASA/CXC/Melissa Weiss; spectrum: NASA/CXC/University of California, Irvine/Taotao Fang et al. Figures 20, 25–26, 29, 35–36, 44, 49, 53–56, 75, 81: NASA/CXC/Melissa Weiss. Figure 22: X-ray: NASA/CXC/University of Hertfordshire/Martin Hardcastle et al.; radio: CSIRO/ATNF/ATCA. Figure 23, 61, 62: NASA/CXC/Sharon Lee. Figure 24: illustration: NASA/CXC/Melissa Weiss; X-ray: NASA/CXC/CfA/Paul Plucinsky et al.; optical: NASA/STScI/SDSU/Jerome Orosz et al. Figures 27, 74: Digitized Sky Survey (DSS). Figure 28: NASA/CXC/Amherst College/Darryl Haggard et al. Figure 30: NASA/CXC/University of Massachusetts/Q. Daniel Wang et al. Figure 31: NASA/STScI. Figure 32: X-ray: NASA/CXC/SAO; UV: NASA/JPL–California Institute of Technology; optical: NASA/STScI; infrared: NASA/JPL–California Institute of Technology; radio: NRAO/AUI/NSF. Figure 33: NASA/

CXC/Kimberly Kowal. Figure 34: X-ray: NASA/CXC/University of Hawaii/Ezequiel Treister et al.; optical: NASA/STScI/Steven Beckwith et al. Figure 37: X-ray: NASA/CXC/CfA/R. Hickox et al.; Moon: NASA/JPL. Figure 38: X-ray: NASA/CXC/SAO/Giuseppina Fabbiano et al.; optical: NASA/ STScI. Figure 39: X-ray: NASA/CXC/SAO/P. Green et al.; optical: Carnegie Observatories/Magellan/ Walter Baade Telescope/J. S. Mulchaey et al. Figure 40: X-ray: NASA/CXC/MIT/Claude Canizares and Michael Nowak; optical: NASA/STScI. Figure 41: NASA/CXC/University of Birmingham/Mark Burke et al. Figure 42: spectrum: NASA/CXC/University of Michigan/Jon Miller et al.; illustration: NASA/CXC/Melissa Weiss. Figure 43: NASA/CXC/Kavli Institute for Particle Astrophysics and Cosmology/Norbert Werner, Evan T. Million, et al. Figure 45: NASA/CXC/IoA/Andrew Fabian et al. Figure 46: SDSS. Figure 47: X-ray: NASA/CXC/SAO/Scott Randall et al.; optical: SDSS. Figure 48: X-ray: NASA/CXC/University of Waterloo/Adrian Vantyghem et al.; optical: NASA/ STScI; radio: NRAO/AUI/NSF. Figure 50: X-ray: NASA/CXC/MIT/M. McDonald et al.; optical: NASA/STScI; radio: Tata Institute of Fundamental Research/Giant Metrewave Radio Telescope. Figures 51–52, 58, 66: NASA/CXC/SAO. Figure 57: NASA/Howard E. Bond and Edmund Nelan (STScI)/Martin Barstow and Matt Burleigh (University of Leicester)/Jay B. Holberg (University of Arizona). Figure 59: NASA/CXC/NCSU/Mary Burkey et al. Figure 60: NASA/STScI/CfA/ Pete Challis. Figure 63: NASA/CXC/PSU/Sangwook Park et al. Figure 64: X-ray: NASA/CXC/ MIT/Laura Lopez et al.; infrared: Palomar Observatory/California Institute of Technology; radio: NRAO/AUI/NSF. Figure 65: X-ray: NASA/CXC/SAO/Fred Seward; optical: NASA/ESA/ASU/ Jeff Hester and Allison Loll; infrared: NASA/JPL–California Institute of Technology/University of Minnesota/Robert Gehrz. Figure 67: Oliver Porth, Serguei Komissarov, and Rony Keppens, "Three-Dimensional Magnetohydrodynamic Simulations of the Crab Nebula," *Monthly Notices of the Royal Astronomical Society* 438: 278, fig. 10. Figure 68: NASA/CXC/University of Toronto/Martin Durant et al. Figure 69: X-ray: NASA/CXC/SAO/Bryan Gaensler et al.; radio: NRAO/AUI/NSF. Figure 70: X-ray: NASA/CXC/ASTRON/B. Stappers et al.; optical: AAO/Joss Bland-Hawthorn and H. Jones. Figure 71: NASA/CXC/MIT/Daniel Dewey et al. and NASA/CXC/SAO/Joseph DePasquale. Figure 72: NASA/CXC/Middlebury College/Frank Winkler. Figure 73: NASA/CXC/Rutgers University/Kristoffer Eriksen et al. Figure 76: NASA/CXC/SAO/Joseph DePasquale. Figure 77: X-ray: NASA/CXC/UAH/M. Sun et al.; optical: NASA/ESA/Hubble Heritage Team (STScI/AURA). Figure 78: X-ray: NASA/CXC/JHU/David Strickland; optical: NASA/ESA/STScI/AURA/Hubble Heritage Team; infrared: NASA/JPL–California Institute of Technology/University of Arizona/ Chad Engelbracht. Figure 79: X-ray: NASA/CXC/MIT/M. McDonald et al.; optical: NASA/STScI. Figure 80: NASA/CXC/PSU/Eric Feigelson, Konstantin Getman, et al. Figure 82: NASA/CXC/ PSU/Leisa Townsley et al. Figure 83: X-ray: NASA/CXC/PSU/Konstantin Getman, Eric Feigelson, Michael Kuhn, and MYStIX team; infrared: NASA/JPL–California Institute of Technology.

RWC
10/18